ROADSIDE
Geology and Biology of
BAJA CALIFORNIA, MEXICO

Second Edition

by
John Minch
Jason Minch

308 Photographs, Diagrams, Line Drawings

jmainc@earthlink.net
JohnMinchBajaBooks.com

John Minch Publishing

Second Edition, First printing © 2017
Printed in the United States of America

Minch, John A
    Roadside Geology and Biology of Baja California / John A., and Jason I. Minch
      Includes bibliographical references
      ISBN: 978-0-9990251-0-9
          1. Travel - Mexico - Baja California, 2. Geology - Mexico - Baja California, 2. Biology - Mexico - Baja California
      I Minch, Minch, Jason I. II Title

Technical Editor: Jennifer Baker

Jennifer Baker has a B.A. in Communications and American Studies from California State University, Fullerton. She is an experienced technical editor, having spent the last 20 years editing deliverables for environmental consultants, transportation planners, civil and structural engineers, and financial professionals. She is also a freelance editor specializing in self-published works.

ACKNOWLEDGMENTS - There are numerous people who contributed to the spirit of this book, both directly and indirectly. We wish to thank all of them for their measure of support. Dave Bloom traveled with us on many trips and contributed much to improving the logs. Particular thanks go to the help and support given by our mother and wife, Carol Minch who put up with the endless hours of work on the manuscript and with the numerous trips into the heart of Baja to compile and check on the information in the logs. Jenifer King (an eighth cousin) made major contributions to the biology of the book in writing difficult sections and providing important information on the birds, plants, and Phytogeographic Regions. Michael Schrauzer worked his magic on the final presentation of this volume.

Our special thanks also goes to Professors Edwin Allison and Gordon Gastil who provided the spark of interest in Baja which led to a burning passion and a lifelong love affair with the peninsula.

TABLE OF CONTENTS

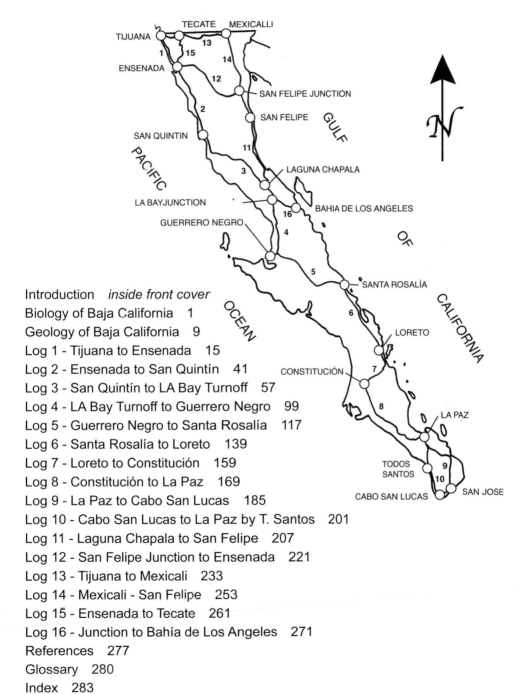

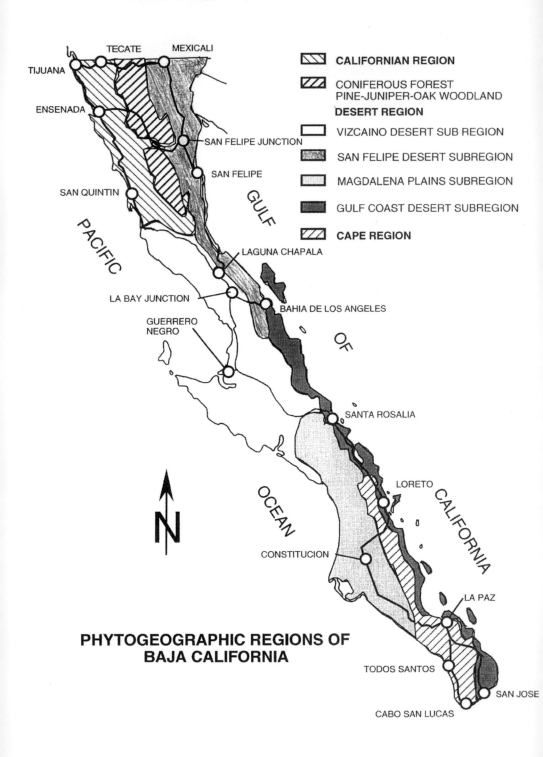

PHYTOGEOGRAPHIC REGIONS OF BAJA CALIFORNIA

BIOLOGY

BOTANICAL WORK ON THE PENINSULA
Much work has been done on the higher vascular plant species of Baja. Naturalist I.G. Voznesenskii conducted plant collecting in Baja as early as 1841. Several excellent botanical collections, botanical accounts, and published references are available that detail the floristics of the vascular plants of Baja for interested travelers, students, naturalists, and serious botanists. References that are particularly noteworthy and useful are by Rebman (2012), Roberts (1989), Standley (1920-26), Shreve and Wiggins (1964), and Wiggins (1980).

PHYTOGEOGRAPHIC AREAS
While traveling along the peninsula, the traveler will notice changes in vegetation. The familiar plants of southern California's chaparral and mountains are replaced by by unfamiliar Desert Region plants. Farther south the equally strange species of the Cape Region appear. These plant ranges are divided into regions or phytogeographic areas. Three major phytogeographic regions of the peninsula are recognized by Shreve and Wiggins (1964), Wiggins (1980), and Roberts (1989).

The **CALIFORNIAN REGION** is an extension of the southern California mountain ranges and their Pacific drainages south to the Sierra San Pedro Martir and some of the mountains in the middle of the peninsula. Vegetative communities of the Californian Region encompass Coniferous Forest, Juniper-Piñon Woodland, Chaparral, Coastal Sage Scrub, and Riparian areas. The Californian Region plant communities and climatic conditions, similar to those of southern California, are found south to San Quintín. Coniferous forests are found in the Sierra Juarez and the Sierra San Pedro Martir. These blend into Pine-Oak Woodlands that are replaced by Chaparral and, near the Pacific, by Coastal Sage Scrub.

Coniferous Forests. Coniferous forests are found along the crests of the Sierra Juarez and the Sierra San Pedro Martir. Common species are Jeffrey Pine, Incense Cedar, Sugar Pine, Lodgepole Pine, Piñon Pine, and White Fir. Other coniferous forest species are Ceanothus, Manzanita, Quaking Aspen, and Canyon Live Oak.

Pine-Juniper-Oak Woodland. Pine-Juniper-Oak woodland is a transition between the pine forests and the chaparral. Common species are Piñon Pine, Coast Live and Scrub Oak, Maguey, Manzanita, Yucca, Penstemon, Rabbitbrush, Sage, Barrel Cactus, and Lilac.

Chaparral. The chaparral covers areas of the western slopes of the mountains south to San Quintín. Chaparral can range from nearly pure Chamise (Chamisal), to one of the most diverse of plant communities.

Chaparral species include Broom Baccharis, Bush Monkey Flower, Ceanothus, Live Oak, Scrub Oak, Flat Top Buckwheat, Indian Paint Brush, Laurel Sumac, Lemonade Berry, Manzanita, Mexican Elderberry, Yucca, Sage, Toyon, and Lilac.

Coastal Sage Scrub. Coastal Sage Scrub is similar to, but distinct from Chaparral, with a more open appearance, and with more Barrel and Hedgehog Cacti. The shrubs are less evergreen, rigid, and woody, with thinner, softer leaves. Many of the species are partially deciduous, dying back during drought periods. Coastal Sage Scrub species include Agave, Barrel Cactus, Sagebrush, Broom Baccharis, Chamise, Cholla, Ceanothus, Dudleya, Fishhook Cactus, Flat-Top Buckwheat, Hedgehog Cactus, Jojoba, Laurel Sumac, Lemonade Berry, Mormon Tea, Mountain Mahogany, Pitaya Agria, Prickly Pear, and Toyon.

Riparian Areas. Riparian Woodlands are vegetational communities that grow along streams and other drainage ways. Since the climatic regime over much of Baja is an arid one, the local occurrence of standing or running water has a dramatic influence on the composition and quantity of vegetation. Localities with permanent standing or running water are generally bordered by deciduous trees, shrubs, and herbs that only grow on the banks of such watercourses. Where river valleys are broad, the Riparian Woodland is correspondingly broad. At higher elevations where the water courses are narrow and the stream banks are steep, Riparian Woodlands may form a very narrow strip that may be only a few feet wide. In the arroyos of the Californian Region, the canopy consists dominantly of Willows with Cottonwoods, Oaks, and Sycamores. The understory is Baccharis, Chamise, Lilac, Scrub Oak, Wild Rose, and Willow.

The **DESERT REGION** comprises the majority of the peninsula, including a narrow strip along the east coast as well as the central and southern regions of the peninsula. The Desert Region is south and east of the Californian Region, and is divided into the San Felipe, Gulf Coast, Vizcaino, and Magdalena Deserts subregions. The climate and vegetation resemble that of the Sonoran Desert, which covers much of the Mexican states of Sonora and Sinaloa, and extends into southeastern California and southern Arizona. Two-thirds of Baja California and all the islands of the Gulf have Sonoran Desert vegetation.

Isolation of the Peninsula has resulted in the development of unusual plants such as the Cardon, Cirio, Elephant Tree, Ocotillo, and Candelillo.

Common plants that occur in one or more areas of the desert region include: Biznaga, California Fan Palm, Candelillo, Cardon, Cholla, Cirio, Datilillo, Desert Mistletoe, Devil's Claw, Dodder, Garambullo, Cactus,

Lomboy, Maguey, Mesquite, Ocotillo, Palo Adan, Palo Verde, Pitaya Agria, Pitaya Dulce, and Torote.

CIRIO - OCOTILLO - CREOSOTE

The Desert Region is commonly divided into four subregions, each with its own dominant species. Many of them occur in more than one subregion. The San Felipe Desert, Gulf Coast Desert, Vizcaino Desert, and Magdalena Plains are all part of the Desert Region.

San Felipe Desert Subregion. The vegetation is very sparse in most areas. Ninety percent of the vegetation in this region is Bursage and Creosote Bush. Other common plants are Brittlebush, Cardon, Cholla,

Desert Willow, Indigo Bush, Ironwood, Mesquite, Ocotillo, Palo Verde, Pitaya Agria, Smoke Tree, Torote, and Yucca.

Gulf Coast Desert Subregion. This region is characterized by trees with swollen trunks including Lomboy and Torote. Brittlebush, Bursage, Cardon, Cholla, Cirio, Creosote Bush, Ironwood, Ocotillo, and Palo Blanco are significant parts of the flora.

Vizcaino Desert Subregion. Most of the dominant plants are succulents or cacti. Though there are Mesquite in many arroyos, trees are scarce. Common plants are Agave, Ball Moss, Burbush, Bursage, Cholla, Cirio, Copalquin, Creosote, Ocotillo, Opuntia, Palo Adan, Pitaya Agria, and Yucca.

Magdalena Plains Subregion. The Magdalena Plains have an abundance of cacti such as Cardon, Cholla, Garambullo, and Pitaya Agria. Mesquite, Palo Adan, Palo Blanco, Palo Verde, and San Miguel are found in the foothills, and Mesquite groves grow in the arroyos. The "fog desert" of this area encourages lichens and Gallitos on the shrubs and cacti. Rocella (*Ramalina* species) is the most noticeable lichen south of Punta Prieta. Other common plants are Datilillo, Lomboy, Maguey, Pitaya Agria, Pitaya Dulce, and Torote.

The **CAPE REGION** includes the Sierra de la Laguna and Sierra de la Giganta and the area south of La Paz. Vegetative communities are Oak-Piñon Woodland and Arid Tropical Forest. The Cape Region is classified as arid tropical forest. Oak-Piñon Woodlands are at higher elevations and a semi-deciduous forest, prominent in leguminous trees and shrubs occupies lower elevations.

Cape Oak-Piñon Woodland. The Oak-Piñon Woodland community is found at higher elevations in the in the Sierra Giganta and the granitic Sierra de La Laguna south of La Paz. There are extensive Mexican Piñon Pine stands in addition to Palmita, Madrono, and Oaks.

Cape Arid Tropical Forest. The arid tropical forest has shrubs and trees of different height and branching character spreading and intermingling with each other, making the region look like an "impoverished tropical jungle," (as described by Shreve). Dominant trees include Mauto, Mesquite, and Palo Verde. In the lower elevations Cardon, Coral Tree, Palo Blanco, Palo de Arco, San Miguel, and Yuca dominate.

BIRDS OF BAJA CALIFORNIA

Birds are seen during the day along the Peninsular Highway. A traveler taking a walk into the different plant communities adjacent to the highway will be rewarded with views of a variety of birds. Although there are some areas that are inhabited by more species than others, some Baja travelers think birding is wonderful throughout the entire peninsula.

The birds of Baja generally exhibit regional habitat preferences and so we find, as with plants, that specific species of birds tend to be found in specific geographical locations. The birds seen along the highway change with the season, geography, latitude, temperature, and weather.

Good sources can be found in Robbins, Brunn, and Zim (1983), Scott (1992), Radamaker (1995), and Peterson's Field Guide to Western Birds.

Radamaker (1995) names 450 species of birds recorded from Baja in his American Bird Association ABA field checklist. The birds of Baja's arid deserts, however, are poorly known, species lists are incomplete, and very few studies concerning the ecology and densities of breeding pairs per acre have been conducted and reported. In the more arid, desert regions of Baja there are fewer birds per acre than in the more dense California region, and those commonly seen are generally drab colored for camouflage and thermo regulation. Most of Baja's plants lose their leaves during the drier parts of the year and consequently birds have little cover to hide in,so they are more visible to birdwatchers.

Four species in particular distinguish the avifauna of the peninsula deserts. They are Scrub Jay, California Quail, Xantus' Hummingbird, and Gray Thrasher. The last two are both endemics of the southern two-thirds of the peninsula and are found in both the desert areas and the Cape Region.

Most of the birds commonly seen along the Peninsular Highway are also commonly seen throughout the Americas so it is difficult to typify the regions of Baja on the basis of its avifauna. An attempt has been made in this guide to list those species most commonly seen along the Peninsular Highway as the traveler passes through the three main phytogeographic regions of the peninsula. Some of Baja's bird species are cosmopolitan over the entire peninsula and so are listed in more than one phytogeographic region.

Several cosmopolitan bird species occur from the International Border to the Cape and on the Gulf islands, in suitable habitats. They are: Costa's Hummingbird, Gila Woodpecker, California Quail, Raven, House Finch, Ladder-backed Woodpecker, Red-tailed Hawk, American Kestrel, , Horned Lark, Verdin, Cactus Wren, Rock Wren, Northern Mockingbird, Blue-gray Gnatcatcher, Black-tailed Gnatcatcher, Hooded Oriole, and Harris' Hawk.

The following information is intended as a brief general introduction to the birds commonly seen along the highway in each phytogeographic region and along Baja's shoreline, as well as Baja's raptors and migratory birds.

CALIFORNIAN PHYTOGEOGRAPHIC REGION:
The bird species of the Pacific coast of Northern Baja are the same as those found in the Californian Region of California. The dominant species of the region are the California Thrasher, Black-shouldered Kite, Anna's Hummingbird, Black-chinned Hummingbird, Red-shafted Flicker, Nuttal's Woodpecker, Scrub Jay, Plain Titmouse, Wrentit, Bell's Vireo, Lawrence's Goldfinch, and Brown Towhee *See* 1:35.5

DESERT PHYTOGEOGRAPHIC REGION:
The dominant bird species of the Desert Region including its four subareas are Gambel's Quail, Vermilion Flycatcher, LeConte's Thrasher, Crissal Thrasher, Abert's Towhee, Gray Thrasher, and Sage Sparrow.

CAPE REGION:
The most commonly encountered species of the Cape Region avifauna (estimated to be 250 species) are the American Robin, Crested Caracara, Northern Cardinal, Pyrrhuloxia, Yellow-billed Cuckoo, Bell's Vireo, Yellow-eyed Junco, Varied Bunting, Common Ground Dove, White-winged Dove, Xantus' Hummingbird, Blue-gray Gnatcatcher, Yellow Warbler, and the Belding's Yellowthroat. Many of the more commonly encountered Cape species are also found in the Desert of the American Southwest. *See* 8:2.

SHORELINE AND GULF ISLAND BIRDS:
The most common shore birds are the American Oystercatcher, Osprey, Gulls, Terns, Pelicans, Egrets, Magnificent Frigate-birds, Blue and Brown-footed Boobies, and the Great Blue Heron. These species are often seen on both the Pacific and Gulf coasts of the peninsula and on virtually every Gulf island. Other birds that are seen on the shoreline include sandpipers, curlews, plovers, avocets, stilts, cormorants, and grebes. Although there are several peninsular endemics, there are no endemic birds on the islands in the Gulf of California. There are other species of shore birds, but they are more patchily distributed in mangrove-fringed beaches and coves on the larger southern islands and along the southern peninsular Gulf coast.

RAPTORS:
Raptors enjoy a wide distribution throughout the peninsula because they lack a habitat preference and are capable of traveling over long distances. The species of raptors most widely distributed in Baja include: Red-tailed Hawks (14:167), Turkey Vultures (8:36), American Kestrels (1:49), Peregrine Falcon (8:34), Northern Harrier (12:143), Ospreys (4:128), and Caracara (9:176). Most of these predatory or scavenging birds are year-round residents of the entire peninsula.

OSPREY IN FLIGHT AT GUERRERO NEGRO

MIGRATORY BIRDS:

Baja lies on the Pacific Flyway, the pathway of many bird species that breed in the summer in western North America and winter in Baja or farther south. As a consequence, a large number of birds, such as Mallards, Brants, teals, and geese, are seasonally seen in Baja. San Quintín Bay is the winter home for the Black Brant and other waterfowl. In the early morning and early evening in Magdalena Bay there are thousands of birds flying to their evening roosts. Low sandy Isla Pajaros looks like there are several birds in every square meter.

ISLA PÁJAROS IN BAHÍA MAGDALENA.

CALIFORNIA BROWN PELICANS AND WESTERN GULL

In each Phytogeographic Region, a list of commonly encountered birds is listed *(Tables 1:27.3, 5:118, 7:103, 8:34, 11:138.7, and 14:167).* At points along the Peninsular Highway, where a particular species of bird has been commonly seen, a brief discussion of the bird, along with a sketch of the bird illustrating its distinctive identifying characteristics is presented.

VULTURE SUNNING WITH BACK TO SUN – WATCHING ME!

GEOLOGY

GEOMORPHIC PROVINCES

Baja California can be divided into five geomorphic provinces (based on geologic landforms). They are: (1) the tilted fault blocks of the *Peninsular Ranges Batholith*, (2) the broad flat plains of the *Cretaceous Geosyncline*, (3) the isolated mountains of the *Coast Ranges*, (4) the fault block mountains and alluviated valleys of the *Basin and Ranges*, and (5) the plateaus of the *Volcanic Tablelands.* The first three provinces are part of the Cretaceous collision of the North American and the Pacific Plate. The last two provinces are directly or indirectly related to the opening of the Gulf of California.

1) The **Peninsular Ranges Batholith** is represented by the granitic ranges of the State of Baja California (such as the Sierra Juarez and the Sierra San Pedro Martir) and by the Cape region mountains (Sierra La Laguna). This province represents the Peninsula Ranges batholith in Baja California. The main rock types are granitic (granite, granodiorite, tonalite, and gabbro), metamorphic (schist, gneiss, and marble), and metavolcanic (metamorphosed volcanic rocks such as andesites, sandstones, and breccias).

2) The **Cretaceous Geosyncline** is largely offshore in Northern Baja with a narrow fringe exposed on land (widest near El Rosario). In Baja California, Sur it underlies nearly all of the state outside of the Cape region. It is generally represented by broad flat areas (Vizcaino and Magdalena Plains) or hidden under the Volcanic Tableland. This province is similar to the Great Valley of California with up to 30,000 feet of sandstone, shale, and conglomerate deposited in a basin in a broad belt in the subduction zone.

3) The **Coast Ranges** are not seen along the highways in Baja California. In northern Baja California it is offshore. In the State of Baja California, Sur it is isolated on Cedros Island, the Vizcaino Peninsula, and the Magdalena Bay islands. Debris from these rocks can be seen in the Rosarito Beach Formation near Tijuana. This province is the Coast Range Province of California. It represents the scrapings from the sea floor of the Pacific plate and mantle rocks (serpentine) that have been metamorphosed and brought back to the surface.

4) The **Basin Ranges** forms the Gulf of California area east of the main Gulf escarpment. Nearly all of the blocky mountain ranges and alluviated valleys of Baja California belong to this province. This province cuts across older provinces. It is based on the presence of block faulting related to the opening of the Gulf of California.

5) The **Volcanic Tableland** forms a broad plateau and mesas, north of the Cape region, in Baja California Sur. These rocks were formed as a result of the passing of the East Pacific rise under the continent and the beginning of the opening of the Gulf of California. They represent extensive outpourings of

volcanic material (tuff andesites, basalts, breccias, and the reworking of this material into volcanic sandstone and conglomerates).

Erosion Surfaces

When the dinosaurs became extinct, Baja California was a relatively flat place with flat to gently rolling topography stretching to the east to the present day Sonora. Remnants of these surfaces dominate the landscapes over which you will drive. Streams are cutting into these surfaces and rapidly destroying them along their edges, however, major pieces of them still exist today (3:127, 12:108, 13:111, 15:32).

Tilted Fault Blocks

The East Pacific rise passed under the continent and began to open the Gulf of California. After some resistance, the continent split, the Gulf opened, and the eastern side of Baja California was arched upward forming the tilted fault blocks of the main ranges. The enormity of these fault blocks is realized when you gently climb for tens of miles along the western side and then steeply descend the eastern escarpments (5:35, 8:34, 13:64).

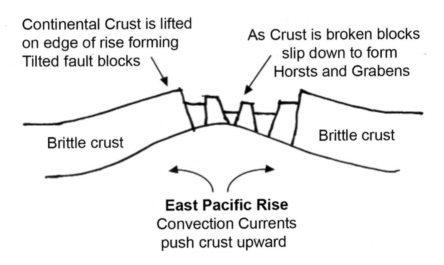

Continental Crust is lifted on edge of rise forming Tilted fault blocks

As Crust is broken blocks slip down to form Horsts and Grabens

Brittle crust

Brittle crust

East Pacific Rise
Convection Currents
push crust upward

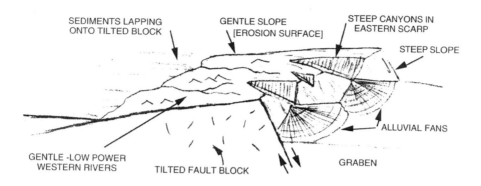

SEDIMENTS LAPPING ONTO TILTED BLOCK

GENTLE SLOPE [EROSION SURFACE]

STEEP CANYONS IN EASTERN SCARP

STEEP SLOPE

ALLUVIAL FANS

GENTLE -LOW POWER WESTERN RIVERS

TILTED FAULT BLOCK

GRABEN

Baja Geologic History

Plate Tectonics

The crust of the earth is a series of semi-rigid plates, moving about relative to each other, and "floating" on the mantle. When the plates pull apart they form Ridges or Rises in Oceanic Crust and rifts [Gulf of California] in Continental Crust. When the plates push together they form Subduction Zones and Island Arcs. When the plates slide past each other they form Lateral [strike-slip] Faults. It is the interaction of the plates that is producing the major features on the earth's crust. The vast majority of this motion is very slow and occurring before our very eyes on a daily basis. A good example is Cajon Pass in southern California. A surveyed point on the railroad through the pass is rising at a steady rate of 16 inches/100 years. That's 16,000 feet in a million years (See 13:64).

Paleozoic - The Quiet Time

The Pre-Mesozoic history of Baja California is obscure, very fragmental, and not well documented. We know that what would later become the North American and Pacific Plates, historically were together as part of a larger plate that was moving eastward, closing the Proto-Atlantic Ocean on its way to a collision with Europe and later, Africa. The west coast was in the middle of this plate like the present eastern coast of North America is today. This stable edge of the continent was receiving clastic sediments from the continental landmass while limestones were being deposited in shallow continental shelf and slope areas. This added a wedge of sediments to the continent. The end of the Paleozoic was marked by the formation of the supercontinent of Pangaea that included all of the continents. In Baja, metamorphosed remnants of Lower Paleozoic carbonates, shales, and sandstones occur in at least one place while Upper Paleozoic rocks have been identified in a number of isolated and scattered areas. Some of the metamorphic gneisses and schists such as the Julian Schist are most likely of Paleozoic age.

Mesozoic - The Big Squeeze

At the beginning of the Mesozoic Era the motions changed. As the supercontinent broke up and the Atlantic opened, North America began to move westward, pushing against the thin oceanic crust of the Pacific part of the plate. The thin edge of the continent buckled and the heavier oceanic crust was forced under the lighter continental crust forming a Subduction Zone and Island Arc with the accompanying volcanoes. Parts of Baja may have been brought from other areas as small landmasses on the East Pacific Plate. However, the majority of the intrusive igneous and metavolcanic rocks in Baja were formed in this subduction zone.

As the Pacific plate was pushed under the continent, the friction caused melting and the formation of magmas. Oceanic crust was continually pushed into and added to the continent in this subduction zone, resulting in uplift and the raising of the magmas closer to the surface. Some of the magmas cooled miles below the surface to form the granitic rocks of the Peninsular Ranges Batholith with its accompanying metamorphic rocks. Other portions of the magma spewed out on the surface and formed volcanoes with the associated volcanic derived sedimentary basins.

Continual subduction and burial of these oceanic crustal rocks caused low-grade metamorphism forming the Alisitos Formation as well as other metavolcanic rocks. As the mountain mass was raising, erosion carried much of the debris westward into the trench basin forming a large Mesozoic sedimentary basin (geosyncline) offshore. The subduction continued for a hundred million years, resulting in the formation of more plutonic rocks after the earlier plutonic rocks were uplifted to the surface and the eroded sediments deposited offshore in the geosynclinal basin and on top of earlier granitic rocks. The fringes of this basin are exposed as the Rosario Formation along the coastline in northern Baja and as a 30,000+ foot thick geosynclinal basin under the majority of southern Baja. The overriding of the East Pacific Rise by the North American plate resulted in the end of the subduction.

Cenozoic - The Big Split and Rip-off

In the Early Cenozoic Era, the peninsula was again a relatively quiet place. The peninsular sierras wore down and a gently rolling erosion surface developed on the exposed batholithic rocks. This surface stretched to the east well into Arizona and Sonora. Major rivers, bearing gravel, flowed across the area from central Arizona to the Pacific.

The North American Plate overrode the East Pacific Rise and began the great rip-off. Coastal California and Baja California began to slide northward along strike slip faults such as the legendary San Andreas Fault in California and the San Miguel Fault, Agua Blanca Fault, Vizcaino Fault, Magdalena Bay Fault, and others in Baja California.

The middle Cenozoic opening of the rift, later to become the Gulf of California, took tens of millions of years. Great sheets of lava and pyroclastic rocks with accompanying volcani-clastic sediments spread over large areas of the peninsula during the Miocene and Pliocene and shallow Miocene seas spread across low areas of the southern part of the peninsula to fill tectonic basins opening in the Proto-Gulf area. The present shape and form of Baja California developed in the last 5-10 million years as the continent finally yielded to the stretching, opening successive areas of the Gulf and then finally, opening the mouth about 5 million years ago. The splitting of the continent tilted the peninsula westward, forming the asymmetric tilted fault blocks of the main ranges of the Sierra Juarez, Sierra San Pedro Martir, Sierra la Giganta as well as other ranges such as the Sierra la Asamblea and Sierra la Victoria.

STYLIZED CROSS-SECTION OF BAJA CALIFORNIA

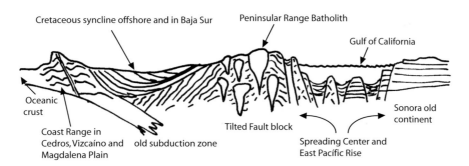

BAJA CALIFORNIA GEOLOGIC TIME SCALE

ERAS	PERIODS	EPOCHS	EVENTS IN BAJA CALIFORNIA
Cenozoic	Quaternary	Holocene	continued faulting and uplift
		Pleistocene	principal uplift and tilting of ranges
	Tertiary	Pliocene	mouth of Gulf of California opens
		Miocene	extensive volcanism - central Gulf opens
		Oligocene	first movements on lateral faults
		Eocene	auriferous rivers flow across the area of the ranges
65 m.a.		Paleocene	shallow coastal seas - tropical weathering
Mesozoic	Cretaceous		subductión zone, formation of Batholith,
	Jurassic		Cretaceous geosyncline, and coast ranges.
230 m.a.	Triassic		Time of Dinosaurs in Baja
Paleozoic	Permian		Volcanism
	Pennsylvanian		shallow seas over much of
	Mississippian		Baja California
	Devonian		
	Silurian		
	Ordovician		oldest rocks in Baja?
600 m.a.	Cambrian		
Precambrian			

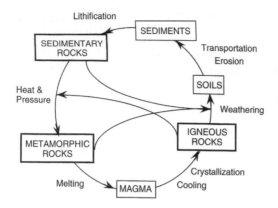

ROCK CYCLE

All rocks are inter-related and part of the cycle of the earth and will eventually go through the complete cycle to molten material. Once rocks are on the surface they tend to stay there and be recycled through the weathering process and erosion.

MAGMA cools and crystallizes to form IGNEOUS ROCKS which weather to form SOILS, etc.

IGNEOUS ROCKS

	Silica Rich		Iron Rich	
Plutonic	GRANITE	GRANODIORITE TONALITE	GABBRO	Coarse Grained- Cooled slowly deep in earth
				◌ Visible Limit
Volcanic	RHYOLITE	ANDESITE	BASALT	Fine Grained Cooled rapidly at or near the surface
	Light		Dark	

Igneous Rocks are formed by crystallization from a magma (molten rock). They either cool slowly, deep in the earth, and are coarse grained. Or, they cool rapidly at the surface and are fine grained. Some (porphyries) cool slowly and then rapidly forming both large and small crystals

OBSIDIAN = volcanic glass
PUMICE =frothy volcanic glass
TUFF = fragments from explosion

SEDIMENTARY ROCKS

		Sediment	Sedimentary Rock	
Clastic (Fragments)		Gravel	Breccia △ Conglomerate ○	2mm
		Sand	Sandstone	1/16mm
		Silt/Clay	Siltstone Shale	
Nonclastic (Chemical)			Chert [SiO_2] Limestone [$CaCO_3$] Gypsum [$CaSO_4$]	

Sediments are formed by weathering and erosion [by water, ice, wind, gravity] of rocks on the earth's surface. Lithification of sediments by cementation, compaction, dessication [drying], and crystallization produces Sedimentary Rocks

BRECCIAS are angular and CONGLOMERATES are rounded fragments larger than 2mm

METAMORPHIC ROCKS

SLATE → SCHIST → GNEISS

Limestone → MARBLE
Sandstone → QUARTZITE
Mantle rock → SERPENTINE
Volcanic rocks→ METAVOLCANIC ROCKS

METAMORPHIC ROCKS are formed by the application of heat and pressure to other rocks. This occurs below the surface of the earth.

Pressure compresses SHALE to SLATE. More heat and pressure causes mica to grow forming SCHIST. As feldspars grow the GNEISS with banding is formed. LIMESTONE becomes MARBLE and SANDSTONE becomes QUARTZITE. SERPENTINE is considered altered mantle material.

METAVOLCANIC ROCKS are metamorphosed volcanic rocks.

Log 1 - Tijuana to Ensenada [110 kms = 68 miles]

The highway climbs west out of the Tijuana River Valley through steep road cuts in the terraces developed on the conglomerates and sandstones of the ancestral Tijuana River delta. It then turns south on the slopes between the high mesas and the low terrace of the Playas de Tijuana to follow the rugged basalt sea cliffs along the elevated Pliocene shoreline with steep rugged canyons, then drops onto a narrow, late Pleistocene terrace cut into the basalt cliffs and finally out into an area of gently rolling basalt and tuff hills. The highway skirts Rosarito on the Pleistocene terrace with views of volcanic capped Mesa Redonda and Cerro Coronel and the rolling volcanic hills, then continues past Punta Descanso where the rolling hills are largely underlain by the marine sedimentary rocks of the Rosario Formation.

By Descanso, the highway is alternately following the narrow late Pleistocene terrace or climbing onto the numerous slump blocks of volcanic rock from the high Miocene basalt capped mesas that form the backdrop for the sea cliffs. The highway crosses the steep sided canyon of the Guadalupe River and travels along a wide segment of the Pleistocene terrace to La Salina before climbing back onto the mesas and slump blocks with the high mesas of resistant basalt overlying the soft marine sedimentary rocks. At San Miguel, the highway follows the Pleistocene terrace past hills of marine sedimentary rocks and then metavolcanic rocks to Ensenada.

Kilometers:

0	International Border: As you cross the border the toll road access is a right turn ramp that is very close to customs. The highway drops into and follows the Tijuana River northward toward the border where the highway bends to the west and parallels the International Border on the right (north).

4	The highway begins to climb steeply into the hills to the west, leaving the flat Tijuana River Valley behind. The highway passes through a series of road cuts that expose conglomerates and sandstones that represent a fluvial part of the Pliocene age San Diego Formation. The ancestral Tijuana River deposited these delta sands and gravels into the San Diego embayment during the Pliocene. Notice that the material of this road cut is stable enough to resist gravity and stands in almost vertical slopes. However, rainwater is making serious cuts in the slopes.

6.3	After the road bends to the south, turn right to stay on Highway 1D. The Tijuana River and the Border are on the right.

6.4	**Los Buenos Fault** - Park at the west end of the turnout at the first bridge over the road. You may need to cross the bridge and make a U-turn to return and park here. Several strands of the Los Buenos Fault are well exposed in the south road cut at and just west of the bridge over the highway. The fault is easily seen on both sides of the road just west of the bridge.

9   Highway 1D curves southward towards Ensenada. A turnoff heading westward from the highway is the main access to the housing development of Playas (beaches) de Tijuana. The Playas de Tijuana is developed on a 55,000-year-old Pleistocene terrace equivalent to the low Nestor Terrace in the San Diego area.

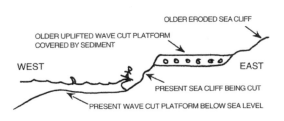

SEA CLIFFS, WAVE CUT BENCHES, MARINE TERRACES, AND BEACHES LANDFORMS OF MARINE EROSION: As waves pound against shorelines, their impact (6,000 lbs./sq.in.) erodes uplifted **sea cliffs** back until a **wave-cut bench** or platform develops at their base. As erosion continues, the wave-cut bench widens until its width absorbs most of the wave energy and a **beach** forms along the now low-energy shoreline. The sea cliff retreats over time due to weathering, erosion, and landslides. In regions of coastal uplift, like Baja, these wave-cut benches, called **marine terraces,** are raised above sea level. Generally, the highway will traverse these marine terraces whenever it is near the Pacific coastline.

There is a good view to the west of Islas Coronados. The toll road (Highway 1D) skirts around the eastern border of the Playas de Tijuana area and continues southward to Ensenada through marine conglomerates and sandstones of Pliocene and Pleistocene age and Miocene basalts.

10   First of three "Casetas de Cobro" (hut of the fees) tollgates.

12.5   The hills to the left are the fossiliferous Pliocene-Pleistocene sandstones and conglomerates of the San Diego Formation.

13   The first exposures of the volcanic and sedimentary rocks of the Rosarito Beach Formation are located near the southern end of the Playas de Tijuana. They were defined by Minch (1967) as a series of basalts (a red-brown to black, fine-grained, iron rich igneous rock), tuffs (a fine-grained fragmental rock of volcanic ash), and tuffaceous sedimentary rocks of Middle Miocene age (16 m.y.) that are exposed between Tijuana and Rosarito.

14   The hills to the left are largely composed of the basalts and tuffs of the Rosarito Beach Formation of Miocene age with a thin veneer of sandstones and conglomerates of the San Diego Formation of Pliocene-

Pleistocene age. The San Diego Formation was deposited in a near-shore environment. The Pliocene shoreline above the highway is a few hundred feet farther inland and a few hundred feet higher than the present day shoreline.

ROSARITO BEACH FORMATION IN SEA CLIFFS

ROSARITO BEACH FORMATION BASALTS OVER TUFFS

14.8   At the La Jolla turnoff, the strata (rock layers) exposed in the road cut

17

are composed of volcanic tuffs overlain by basalts. These tuffs and basalts are part of the Costa Azul Member of the Rosarito Beach Formation. A prominent bake zone of reddish tuff is exposed in the outcrop. It formed in the Middle Miocene as molten basalts flowed over the white tuffaceous material and baked them to extremely hot (2000° F) temperatures.

The La Jolla turnoff provides an opportunity to observe the lithologies and relations between two members of the Middle Miocene Rosarito Beach Formation and the Upper Pliocene San Diego Formation.

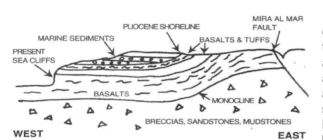

Franciscan-type detritus: The base of the section is located about a mile up the arroyo to the east around the southeast side of the ridge. Exposed here is the lowest, Mira al Mar Member, of the Rosarito Beach Formation. This member is exposed in the core of a monocline and represents mudflows composed of a light-gray, medium to coarse-grained, arkosic, sandy matrix breccia. These breccias contain fragments of Franciscan-type detritus such as glaucophane schist and other schists, serpentine, sausuritized gabbro, bedded chert (a dense, hard, siliceous rock), quartzites, and minor amounts of acidic, volcanic, and plutonic rocks (Minch, 1967). In another canyon, an oolitic limestone (spheroidal or ellipsoidal particles formed by chemical precipitation in shallow, wave-agitated water) and tuffaceous sandstone are exposed in the section indicating a relatively quiet deposition between the lobes of this submarine mudflow fan complex.

This member has been interpreted to represent a series of mudflow deposits originating from a western landmass and deposited on a narrow shelf against and onto the slope of the landmass area to the east. Between mudflows, the ocean currents and wave action reworked the mudflows that produced sandy matrix breccias next to the lobes, and sandstones and shales more distant to the lobes. This area, between the offshore volcanic highlands and the Baja California mainland, was periodically swept by strong currents. This unit is exposed along this arroyo and in canyons along the Los Buenos Fault in the Tijuana-Rosarito coastal area (Minch, *et al.*, 1984). This member is correlated with the Los Indios Member exposed to the south in the La Misión area.

A short walk along the dirt road and to the left leads to a fossil locality in the near shore Pliocene San Diego Formation that overlies the basalts and tuffs of the Coastal Azul Member. The fossils occurring at this site (and in another notable locality in the hills east of Km 15.2) are found in poorly sorted yellow-brown conglomeratic sandstone beds resting directly on the weathered surface of the eastward-sloping Miocene basalts. Someone has stripped this outcrop of all traces of fossils.

Do not collect Baja's fossil materials. In addition to it being against Mexican federal laws to collect any of the peninsula's resources, the loss of these materials leads to the loss of the ability to illuminate Baja's past.

15.2   A richly fossiliferous San Diego Formation locality on the north side of this canyon just east of the highway yielded an invertebrate fauna of 36 species from lensing sandstone beds about 6 feet above the contact with Miocene volcanic rocks.

A MARINE PLIOCENE SHORELINE AND FOSSILS: The rich fossil fauna of the two localities (Kms 14.8 and 15.2) is dominated by shells and shell fragments of the extinct molluscs: *Pecten healeyi*, *Acanthina emersoni*, *Anadara trilineata* and *Chalmys parmeleei*. These extinct fossil species are characteristic of Pliocene strata found on both sides of the International Border.

The environmental living conditions of both the extinct and extant (living) species of mollusks in this region was one in which the ocean temperatures were cooler than those presently encountered in the San Diego-Tijuana region.

The fossil fauna at the two localities can be divided into two cool water components. The first is a **littoral** and **inner sublittoral epifauna** that lives on an exposed rocky substrate. This component is represented by the indicator genera: *Acanthina*, *Calliostoma*, *Balanophillia*, *Penitella*, *Tegula*, and *Thais*.

The second component consists primarily of the following **sublittoral** indicator genera: *Acila*, *Dentallium*, *Dosinia*, *Laevicardium*, *Nuculana*, *Panope*, *Protothaca*, *Siliqua*, *Spisula*, *Terebra*, and *Tresus*. The environmental requirements of these genera suggest a sublittoral, semi protected silt or sand substrate environment. The mollusc *Calyptraea mammilaris*, a warm water organism, indicates that a warm Infaunal element also existed as a subcomponent of this cool water environmental element.

Also present at this locality is a substantial vertebrate fauna that includes the large teeth of *Carcharadon megalodon*, the huge extinct cousin to the modern Great White Shark (*Carcharadon carcharias*). *Carcharadon sulcidens*, another Great White Shark, is also present.

Other vertebrates present at this locality include whale and dolphin remains, a Mako shark (*Isurus planus*), a bay shark (*Carcharias sp.*), and a bay ray (*Myliobatis sp.*). *See* Ashby and Minch (1984) and Rowland (1972) for more detailed discussions of the paleontology at this locality.

15.5   The upper part of the road cut on the left exposes a series of prominent cross-bedded Pliocene sandstones. These overlie Miocene basalts of the Rosarito Beach Formation. Recent data (Luyendyk et al., 1988) from the Rosarito Beach Formation flows were collected east of the road and dated by $^{40}Ar/^{39}Ar$. A date of 16.2 +/- 0.3 Ma was reported for the basal basalt; and 16.14 +/- 0.06 on the overlying upper olivine basalt of the Amado Nervo Member. These dates seem reasonably close to the 15 m.y. ages for basaltic vulcanism on the Continental Borderland and are probably representatives of a volcanic province that was active in mid-Tertiary time (Hawkins, 1970).

16.1   From the Punta Bandera turnoff the highway follows a narrow strip of the Late Pleistocene Terrace all the way to Rosarito. From this point good exposures of the Miocene fluvial sedimentary rocks are visible on the middle rock of the north island and the south island of Islas Coronados.

A short walk north along the beach at this turnoff reaches excellent exposures of airfall lapilli tuffs and basalts in the Costa Azul member of the Rosarito Beach Formation.

Fossils from this well-developed Pleistocene terrace consist of shallow water marine invertebrates mixed with aboriginal and land mammal remains.

In this region the Nestor Terrace was very heavily occupied by several Indian cultures. Evidence of prior habitation is found in numerous areas along the terrace in the form of "kitchen or trash middens".

INDIAN KITCHEN OR TRASH MIDDENS: On the north side of Cañon San Antonio de Los Buenos, an extensive Indian trash midden is exposed in the cliff just above the highway. The upper dark soil layer represents remains of aboriginal occupation of the terrace. For many years native Indians lived along the Pacific coastline. They did not have permanent villages, but lived together in small family camps that they moved when the area was "fished out" or became infested with fleas (pulgas). As the camps moved up and down the coast an almost continuous layer of midden material was deposited over the years. Some anthropologists believe it is possible that Indians may have occupied this particular area as early as four thousand years ago. Excavations at midden sites like these are very valuable to archaeologists and anthropologists; they provide many man-made artifacts, burials, and plant and animal organic remains that help to reconstruct their life-style.

Do not disturb the midden or remove anything from Baja. It is ILLEGAL! *See 3:176* for a discussion of the prehistory of early man in Baja.)

19.0   The highway crosses Cañon San Antonio de Los Buenos. In 2017, the Tijuana sewer still utilized this canyon as an access to the ocean. Three Kilometers up this canyon, the Mira al Mar Member of the Rosarito Beach Formation contains Franciscan detritus (*See* 1:14.8) and Miocene fossils.

22.0   This Late Pleistocene Terrace near San Antonio Shores has also experienced severe sea cliff erosion during the last three decades.

Excavations on the Nestor terrace at San Antonio shores have yielded the remains of a possible new species of mastodon (*Stegamastodon sp.*), a relatively recent relative of the modern horse (*Equus caballus*), and other mammal bone fossils. These fossils represent a land fauna that occupied this low coastal terrace following its late Pleistocene emergence.

22.5   The road cuts for the next 11 Kilometers exposing basalts and tuffs of the Rosarito Beach Formation. The reddish tuffs near the top of this section were baked by the Miocene basalts (*See* 1:14.8).

26   The highway bends to the left. The Rosarito Power Plant to the right burns fossil fuels to supply much of the electricity for northwestern Baja California. Tankers delivering fuel to the plant are often seen moored offshore.

To the east is the prominent steep-sided flat-topped Mesa Redonda. The prominent peak to the right of Mesa Redonda is called Cerro Coronel. Both hills are capped by Miocene basalts and are underlain by Eocene and Upper Cretaceous strata. Mesa Redonda was formed as the underlying sedimentary rocks were eroded by streams that dissected the older landscape. This left the flat-topped mesas standing above more erodible areas.

BASALT CAPPED MESA REDONDA

27.3   The highway descends a fault scarp as it crosses the active Agua Caliente Fault that offsets the Late Pleistocene terrace in this area. This fault was given this name because it runs through the Agua Caliente Race Track and is responsible for the Agua Caliente Hot Springs at the racetrack. The Agua Caliente Hot Springs is one of a number of hot springs located along faults in Baja California (*See* 13:4.4). This fault also runs close to the area of the Rosarito Power Plant and tank farm.

THE CALIFORNIAN REGION AVIFAUNA: The bird species of the Pacific coast of Northern Baja are the same as those found along the Pacific shoreline and in the Chaparral and Oak-Woodland plant communities of the Californian Phytogeographic Region of southern California.

Where a species of bird has been frequently seen along the highway, a brief natural history discussion will be presented for that bird along with a line sketch and a discussion of distinctive identifying characteristics.

BIRD NAME	LIKELY LOCATION
Birds found throughout Baja in suitable habitats:	
American Kestrel	Wires and fence posts
American White Pelican	Gliding along the shore
Anna's Hummingbird	Feeding on red or yellow tubular flowers
Ash-throated Flycatcher	Desert, chaparral, woodlands
Black-chinned Hummingbird	Feeding on red or yellow tubular flowers
Black-tailed Gnatcatcher	Flying, flitting through low chaparral brush
Blue-gray Gnatcatcher	Flying, flitting through low chaparral brush
Cactus Wren	On cacti
California Brown Pelican	Gliding along the shore
California Quail	On the ground
Common Raven	Flying and perching on trees, fences, poles
Greater Roadrunner	Crossing the highway
Hooded Oriole	Common around palms
Horned Lark	Dirt fields, gravel ridges, and shorelines
House Finch	Abundant in all areas
Ladder-backed Woodpecker	Flitting in the air
Northern Mockingbird	Thickets, woodlands, towns
Red-Tailed Hawk	Tops of telephone poles and fence posts
Rock Wren	Scrublands, dry washes, & most arid areas
Turkey Vultures	Soaring or on carrion on highway
Verdin	Mesquite and thorny shrubs
Western Meadowlark	Fence posts and wires

BIRD NAME	LIKELY LOCATION
Birds of the California Region:	
Bell's Vireo	In moist woodlands, bottomlands, mesquite
Brown Towhee	Brushy hillsides, wooded cyns, & chaparral
California Thrasher	On the ground
Killdeer	edges of marshes, shorelines, and salt flats
Lawrence's Goldfinch	In the chaparral and dry, grassy areas
Loggerhead Shrike	Wires and fence posts
Nutgall's Woodpecker	Chaparral, oak woodlands
Plain Titmouse	On oaks, junipers, pines
Red Shafted Flicker	Chaparral oak woodlands
Scrub Jay	Chaparral oak-woodlands
Starling	Flocking, wires, trees, grain fields
Wrentit	Chaparral, coniferous woodlands

The **CALIFORNIAN PHYTOGEOGRAPHIC REGION** dominates northern Baja. There are five vegetational groups in this area: the Coniferous Forest, Piñon-Juniper Woodland, Chaparral, Coastal Sage Scrub, and Riparian areas. Each is typically characterized and recognized by its association of dominant plant species known as indicator species. *See the Biology Chapter for descriptions of the Phytogeographic Regions of the Baja Peninsula.*

27.7   Excavations in a road metal quarry to the north of the highway have yielded a diverse and abundant Late Pleistocene assemblage of marine invertebrates. White sandstones high on the cut at the northeastern end of the quarry near the crest of the hill into which the quarry is excavated yielded 55 species of invertebrates at two localities. No southern faunal forms have been recognized. A conglomerate on the floor of the southeast face yielded only 7 species and includes the extinct mollusc *Crepidula princeps*.

29.4   The Rosarito turnoff is a business loop through the town of Rosarito that eventually returns to the toll road near Km 35.

29.8   The free road from Tijuana joins the highway. The highway continues south on the Late Pleistocene terrace cut on the Miocene volcanic sequence.

33.6   Just south of town is the long-standing Renee's Rosarito Beach Motel for which the Rosarito Beach Formation was named.

34.5   The road from Rosarito returns to the toll highway where the free road leaves the highway. The free road parallels the highway to La Misión and provides access to local beaches. This road log (guide) follows the toll road.

35.5   Caseta de Cobro

36   South of the toll station, the highway follows a Pleistocene terrace developed on the Rosarito Beach and Rosario Formations that are locally interrupted by landslides where weak rocks underlie the capping basalts.

39   There is a prominent hill to the left of the highway with what looks like vertical columns or posts. It is the result of a geological phenomenon known as columnar jointing. This hill is the neck (volcanic plug) of a volcano younger in age than the Rosarito Beach Formation. It is one of a number of volcanic plugs that dot the International Border area from Tijuana to the east for tens of kilometers.

VOLCANIC PLUGS are masses of rock that seal the vents and conduits of volcanoes. These vents become exposed as the more erodible surrounding rock of the original cone is removed.

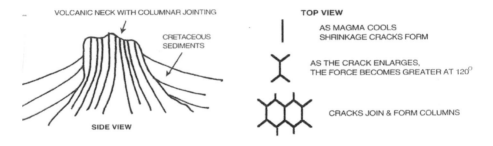

40   Baja Studios. The movie TITANIC was filmed at this location. The junction with the new highway to eastern Tijuana, Otay, and Tecate heads east at this point. Not accessible from toll road if traveling southbound.

42   The outcrops in the hills are the marine sedimentary beds of the Cretaceous Rosario Formation.

This Pleistocene terrace was once more extensively developed. As a result of erosion of the softer sedimentary rocks, the terrace has been reduced to remnants with intervening areas weathered into rounded hills and gullies.

Late Pleistocene marine terrace deposits rest on Miocene volcanic rocks here and elsewhere along the local coastline. On the west side of Punta Descanso just north of the point, about 6 feet of fossiliferous material is exposed adjacent to the cut terrace surface along about 120 feet of sea cliff. The lower 3 feet of that deposit is an unconsolidated pebble conglomerate in a sand matrix. The upper 3 feet is unconsolidated rubble with a matrix of shells. These two rock layers have yielded about 150 invertebrate species. Thry are chiefly molluscs, including *Chione picta* (living from Bahía Magdalena southward) and *Velutina laevigata* (living north from Cayucos, California).

44.5   An Indian midden is exposed in the road cuts on both sides of the road for the next 0.5 kilometer. If left undisturbed, these middens will provide valuable information about the history of man in Baja as future archaeological studies are conducted.

45   The basalts of the Rosarito Beach Formation overlying the Rosario Formation are well exposed in some of the cliffs east of the highway.

46.5   For the next four kilometers there are good views to the east of Cerro Coronel. The small basalt mass offshore is known as El Moro (snout) Rock.

48   For the next kilometer, there are good exposures of the Rosario Formation in road cuts along both sides of the highway and in the sea cliffs.

49   This is the turnoff to the communities of Cantiles and Puerto Nuevo.

The poles and wires in this region are usually good places to look for American Kestrels.

American Kestrel is the smallest and most vocal among Baja's birds of prey and is the most commonly seen falcon of the open country in Baja. This is also the only falcon with a rusty back and tail. It hunts from poles, wires, or trees and is seen hovering in the air before it stoops to capture some ground dwelling insects, reptiles, or rodents.

52.2   The highway crosses Valle El Moro. The bridge over the highway provides an unpaved access to the valley from the old road.

California Quail: They are members of the pheasant subfamily and are small plump-bodied intricately colored birds that live in the coastal and foothill chaparral, live-oak canyons, deserts, and oasis throughout Baja. They often travel and feed in large coveys. Their most identifiable field characteristics are their bobbing black topknot, feathers, and scaled chests. Depending on the location of the accent the voice of these shy skittish birds seems to say "*come* right here" or "where *are* you?"

54.5   The Medanos coastal dune field formed as a result of ocean sands blown on shore by strong winds coming from the open sea. The dunes are stabilized (they don't migrate) by the roots of the vegetation that grows on them. Sometimes the dunes can be "blown out" by wind or exceptionally high waves. If this occurs, they start to move and are no longer stabilized. When the dunes are "blown out" a secondary type of plant succession known as "old dune succession" occurs as vegetation returns to stabilize the dunes *See* 3:21, Sand Dunes in Baja.

MEDANOS DUNE FIELD

55.5   For the next 40 Kilometers between here and San Miguel the highway will passing between the ocean and the high mesas of the La Misión Member of the Rosarito Beach Formation,.

BASALT CLIFFS IN DESCANSO VALLEY

58.6   Cuenca Lechera turnoff provides access to Medio Camino.

59.4   Medio Camino (Half Way House) is located halfway between Tijuana and Ensenada. The Half Way House is built on tuffs and basalts of the Rosarito Beach Formation. The mesas visible to the east of the highway are capped by the Rosarito Beach Formation that overlies the Cretaceous Rosario Formation. The gently dipping tuffs of the Punta Mesquite Member of the Rosarito Beach Formation outcrop in the sea cliffs north and south of this point. The lithic tuffaceous sandstones beneath the Half Way House are slightly different in composition from the tuffs to the north. They are coarser and contain interbeds of siltstones and shales and have been highly faulted. A walk along the beach to the north reveals exposures of this tuff.

Examples of sedimentary structures to be examined include: channel cross-bedding, hummocky cross-stratification, channel cross-stratification, load structures, rip-up clasts, and bioturbation (*Omphiomorpha*) of various kinds.

This area is interpreted as proximal to the offshore volcanic highlands during the Miocene. Deposition of these Miocene sedimentary rocks was above storm wave base (Hummocky cross-stratification) and well oxygenated (bioturbation). The sedimentary structures all exhibit an eastward transport direction. This area contains the oldest exposures of the Rosarito Beach Formation. Preliminary findings indicate that these tuff exposures represent a portion of the western margin of the basin during the Middle Miocene.

62   To the north is a panoramic view of the coastline with Punta Descanso in the distance. Tuffs of the Rosarito Beach Formation are exposed in the sea cliff. On a clear day, the southern island of Islas Coronado is visible offshore.

64   Plaza del Mar with a replica of a Mayan Pyramid was built in the middle of a massive landslide bowl. This area exhibits a low hilly (hummocky) topography and indicates that many landslides have occurred here. In the distant past the whole coastline has slid downward and westward. Locally, prominent terraces have developed on the La Misión and older members of the Rosarito Beach Formation that attests to the relative stability of some of these landslides. Many of them occurred in excess of 50,000 years ago.

SLUMP BLOCKS are common along the highway for the next 3 Kilometers. The closed depressions are unaltered by drainage. Many slumps project into the sea like brown gnarled fingers.

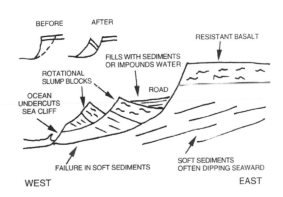

Coastline slump block failures: South of this area the toll highway crosses Quaternary to Recent aged landslides for approximately 40 kilometers along the rugged coastline ranging up to 500 feet above sea level. Most of the sea cliffs are underlain by a thick sequence of flat to gently southwest-dipping upper Cretaceous, conglomerates sandstones and mudstones of the Rosario Formation. These softer and more easily erodible sedimentary layers are overlain by resistant Miocene basalts of the Rosarito Beach Formation.

66   The La Fonda turnoff is the last turnoff for traffic between the toll road and the free road. The toll road is the shorter route to Ensenada. The La Fonda Resort is a popular tourist area situated on a basalt Sea cliff on the very narrow Pleistocene coastal terrace. Demetri (Demetri's at South end) has been one of my favorite hosts and his terrace dining is a must. Dolphins can be seen in the surf and pelicans can be seen skimming close to the waters just offshore. Beware of the pelicans as I've had them fly by less than 10 feet above my head.

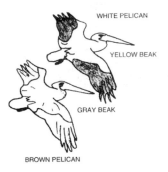

The American White Pelican and the California Brown Pelican are the large aquatic fish-eating birds seen along both coasts and on many of the gulf islands in Baja where they nest in large colonies. They are commonly seen flying in long straight lines (brown pelican) or in V- shaped formations (white pelican) only inches above the surface of the water. Feeding is accomplished by diving into the sea from as high as 150 feet (brown pelican) or by bill scooping as they wade or swim in shallow coastal or lake waters (white pelican). The slow deliberate flight of the brown pelican, low over the water with sudden plunges for fish, makes its identity unmistakable. The brown pelican is commonly on fishing piers and around docks.

BEACH AT LA MISIÓN

67.5   A view of the older development of La Misión reveals a fine flat beach and a scenic rocky headland. At one time this was the only developed area along the highway between Tijuana and Ensenada. The road cuts in this area are the basalts of the La Misión Member of the Rosarito Beach Formation.

69.2   The highway crosses the estuary of the Guadalupe River that originates in the high Sierra de Juárez near Laguna Hanson. Deep wells in this valley provide part of the water supply for the cities of Tijuana and Ensenada.

**GUADALUPE RIVER AT LA MISIÓN
WITH SALT MARSH AND BASALT MESAS**

Coastal Wetlands are **Estuarine Environments** that exist where the ocean tides meet the outflow of a river current. The water of this environment may be alternately fresh or brackish. The marsh and aquatic plants are of great value to shorebirds, to some small mammals, and to songbirds such as the Red-winged Blackbird, Sparrow, and Marsh Wren. The characteristic plants in this environment are members of the rush family. Spiny Rushes (*Juncus acutus*) are tufted grass-like herbs commonly found growing in moist places like the Guadalupe Estuarine Marsh. The durable stems of this herb were used by early man in the construction of baskets. Other plants of this marsh are Yerba Mansa, Salt Grass, Alkali Heath, Heliotrope, Sea Lavender, Salt Cedar, Glasswort, Pickleweed, Sea Purslane, and Sea Blite.

The birds most often seen here are the Killdeer, Red-winged Blackbird, Egret, and Great Blue Heron.

29

The **Killdeer** is widespread throughout the entire peninsula. Because of its plaintive lonely cry, *kee-kil-de-dee*, early Spaniards named this bird El Perdido, "the lonely one." It is often seen and/or heard in agricultural fields, short grassy areas, and along the borders of salt marshes and the shorelines of both of Baja's coasts. The distinctive identifying characteristics of the Killdeer are its double black-  banded white neck and orange upper tail and lower back. It performs a "broken wing routine" when nests or young are threatened. Their open depression-like nests are usually built on gravelly soils.

The Red-Winged Blackbird is a resident of fresh water marshes, fields, and moist grasslands in northern Baja. The males are easily recognizable by their red shoulder epaulettes. The epaulettes are used to defend their territory from other male Red-Wings. This abundant, aggressive species is often found in immense flocks in winter.

The **Great Blue Heron** is often seen in the marshes or on the beaches, or even in the dry fields of Baja. The tall, lean solitary figure of the Great Blue Heron can be seen standing motionless in a pool of water or advancing slowly one step at a time, lifting each foot stealthily from the shallows without a ripple. Herons may stand as still as a statue for over half an hour while waiting for prey. With a lightning quick forward  lunge of their long neck and bill, the heron captures its prey. They prefer fish but will also eat birds, small mammals, insects, snakes, frogs, and crustaceans.

Whether it is on land or in the air, a Great Blue Heron is recognized by its yellowish bill and ornate plumes on its head, neck, and back, the long snakelike neck in an "S" shaped curve, the slow wing strokes, long legs, and 6-foot wing span. In Baja, look for the solitary Great Blue Heron on piers, docks, sandbars, or in estuaries, coves, marshes, and the riparian woodland.

The **Great (Common) Egret** resides throughout the lowlands of Baja near fresh water streams, salt marshes, ponds, mudflats, and estuarine environments. It is one of the larger wading birds in Baja, larger than any other heron except the Great Blue Heron. The Great Egret has white plumage, a yellow bill, and shiny black legs and feet. Like the Great Blue Heron, the Great Egret is a slow moving patient hunter of shallow water fish.

69.5   An excellent turnoff alongside the highway provides easy access to a wide and flat beach. Many beaches form where there is an abundant supply of sand, such as the Guadalupe River. Waves distribute them along the coastline.

70   The basalt forming the road cuts here and on which the coastal homes in the area are built may represent a massive slump block or a fault block. An alternate theory is that this basalt was deposited against the Cretaceous sea cliff and is older than the basalt that forms the mesa top. The overlying tuffs were stripped from much of the coastal area which left a flat surface along the coast for several kilometers.   The Late Pleistocene Terrace (50,000 years old) has developed on top of this and shows the antiquity and stability of some of the coastline between Tijuana and Ensenada. Farther south, the coastline is highly unstable and is actively moving downward and seaward. This is demonstrated by the numerous repairs that are continually being made on the undulating "bumpy" highway.

72   On the right is Baja Ensenada RV Trailer Park. Note the super-tidal area. During some future particularly violent Pacific storms, part of this park may be inundated by high tides.

The Nopales cacti growing along the highway on the left were planted for use as food.

THE NOPAL CACTUS (*Opuntia fincus indica*) belongs to the family Cactaceae and looks like beavertail cactus. Young Nopal leaves are larger and do not develop spines right away. They have little knobs that are not sharp. Nopales leaves are harvested young, shredded, and used to add fiber and bulk to the diet. They are cooked like green beans, fire roasted, or pickled with chilies.

73   The tidal flat (salt marsh) of La Salina just south of the turnoff is also a super-tidal area. Frequently, the mouths of many of Baja's coastal river valleys are semi-sealed by a sand bar (stretches of beach dunes that separate the open sea front from the tidal flats to the rear.) This provides for the tidal flat with occasional access to the sea. They have built a marina entrance through the bar at the mouth of this marsh.

At one time, this tidal flat must have been very popular with the Indians, as fairly extensive Indian kitchen midden (trash) materials extend eastward from the highway for about a kilometer on both sides of the marsh. The highway turnoff cuts through and exposes a small portion of this vast midden. A very large (12 cm.) perfectly worked spear point of black metavolcanic material was discovered in the triangle of ground formed by the two roads that enter and exit the highway. *(Refer to 3:176 for history of man in Baja.)*

LA SALINA ESTUARY IN EARLY 2016.

TIDAL (MARSH OR MUD) FLAT ENVIRONMENTS: In general, Baja's tidal flats occur between mean high tide and mean low tide levels, are vegetated by unicellular and larger forms of algae, and are bordered on their inland edge by intertidal Pickleweed and other halophytes (salt loving plants). Intertidal Pickleweed begins its best growth at the average high tide line.

Tidal flats are the home or favored resting place for many living organisms adapted to life in this extremely harsh saline environment. Organisms that inhabit tidal flats face severe changes in daily environments. They are alternately immersed by high tides or exposed by low tides. Part of the tidal flat will always be underwater in sloughs and channels, and part will be above the high water mark. Because of these changing conditions, variations in soil salinity are to be expected.

In the lowest portion of the mud flats the salt concentrations may exceed 6% preventing the growth of all plants except marine algae. Near the shallows and edges of the flats, salinity both decreases and fluctuates with fresh water runoff and rain. Temperature ranges are broad which necessitates adaptations for surviving fluctuations of as much as 50° F in a single day's cycle.

Closer to the shore, away from the shallows and on the banks bordering the tidal flat, halophytic plant roots capture silt, mud, and sand. The halophytes growing around La Salina marsh are divided into two communities based on their relationship to the tidal flat. The **DUNE AND STRAND LINE FLORA** are predominated by halophytic Sand Verbena, Sand Bur, Beach Evening Primrose, Beach Fig, Door Brush, and Ice Plant; and the **SALT-MARSH FLORA** are predominated by halophytic Yerba Mansa, Salt Grass, Alkali Heath, Heliotrope, Spiny Rush, Sea Lavender, Salt Cedar, Glasswort, Pickleweed, Sea Purslane, and Sea Blite.

Farther inland as the land slopes up to the surrounding hills covered by true Coastal Sage Scrub, the more salt-tolerant plants are replaced by those that are less tolerant.

Although they are not very obvious, animals thrive in mud and tidal flats (salt marshes). Due to increased nutrient flows, tidal flats are very fertile and are more prosperous than many other natural communities. Tidal flats are busy places especially for birds such as Pintails, Mallards, Coots, Gulls, Terns, Curlews, Willets, Dowitchers, Yellow Legs, and Whimbrels that open mollusk shells and search for worms and shrimp in the tidal flats.

76   The highway passes through a road cut in the tuffs and tuffaceous sandstones of the Rosarito Beach Formation. In this area, the tuffaceous strata of the Rosarito Beach Formation has yielded a Middle Miocene marine fossil fauna that includes: *Chione temblorensis*, *Anadara topangansis*, *Turritella ocoyana*, and marine vertebrates. These strata have also yielded a camelid that has been tentatively identified as *Oxydactylus* cf. *longpipes*. This combination of marine and nonmarine species in the same locality suggests a marine-nonmarine tie-in between the Middle Miocene Hemingfordian North American land mammal age and the West Coast Temblor marine molluscan stage (Minch *et al.*, 1970). The mesa top basalt has been dated at 16.5 million years.

77.8   The Baja Mar golf course has been described by many players as being "challenging" because the fairways are "islands" in a "sea" of Cactus, Agave, and Coastal Sage Scrub. It has been said that errant players are just as likely to find their ball impaled on a cactus as resting on the fairway.

78   Just south of Jatay, the hills to the left of the highway seem to be quite straight (linear in relation to the coast) and have a concave upward slope. This concave upward slope is the headwall of a very large rotational slump block. The land that this portion of the highway is on has slumped down and rotated. All along the highway there are numerous concave areas that are indicative of the rotation of numerous such smaller slump blocks.

84   The turnoff to El Mirador is on a dangerous curve. The edge of a 1,000-foot escarpment at El Mirador provides a view of coastal features that are visible in this area and include a view of Islas de Todos Santos.

The panoramic view of the coast to the south shows the mudstones, sandstones, and conglomerates of the upper Cretaceous Rosario Formation overlain by the basalts of the Rosarito Beach Formation. Numerous slumps are evident along the coast. Some have dropped capping volcanic rocks to sea level where they now form resistant shore promontories. The metavolcanic spine of Punta Banda on the south side of Bahía Ensenada is often visible in the distance. It is a fault block bound on the northeast and southwest sides by two branches of the Agua Blanca Fault. The Islas de Todos Santos are the offshore Islands

**COASTAL VIEW WITH TERRACES
ON SLUMP BLOCKS BELOW HIGHWAY**

ISLAS DE TODOS SANTOS: The two islands of Islas de Todos Santos are composed of middle Cretaceous metavolcanic rocks of the Alisitos Formation overlain by basalts (La Misión Member) and sedimentary rocks (Los Indios Member) of the Rosarito Beach Formation. Also present on the islands is a well-developed, Late Pleistocene marine terrace. The waves breaking on the north end are reported to be among the highest in the world.

85 If you missed the turnoff to El Mirador at Km 84, this turnout has a view similar to that at El Mirador. This spot provides a good place to look at the flat terraces below the highway. Each terrace is a rotational slump block. The viewpoint produced by the rotational movement of a slump block is directly above one of these terraces.

SLIPPING AND SLIDING: Because much of the coast is slumping into the Pacific along this stretch of the highway, the pavement from here to San Miguel undulates considerably. Between Salsipuedes and San Miguel, active sliding caused by marine erosion requires continuous highway maintenance to compensate for the seaward movements of large slide blocks. Portions of the highway have been down-dropped several feet. New pavement marks the lateral edge of several slide blocks, and fresh scarps several feet high in natural ground mark the up-slope boundary of active sliding. Preventative and corrective road work in progress includes unloading of the head of the slide, draining groundwater from within and beneath the slide, and rechanneling surface drainage to prevent further infiltration of water into the slide.

Watch for Brown Pelicans that soar along on sea cliff thermals between here and Ensenada. The mourning dove is also common in this area.

Mourning Doves: These are small slim birds with long sharply-tapered tails and black spots on the upper wing. They are commonly seen in northern Baja in gleaning seeds among the stubble of cultivated grain fields in the late summer or fall. They are frequently seen in cattle grazing ranges on dry uplands and in many of the villages and desert areas of Baja. Except for an occasional American Kestrel, doves are the largest birds that commonly perch on telephone wires where their long sharply-pointed tails make them  easily identifiable. As you reach the southern part of the peninsula, watch for a close relative, the White-Winged Dove.

87.5   Salsipuedes: In Spanish aptly means "Get out if you can."

91   From here to Km 94, there are exposures of the massive sandstone and conglomerate lenses of the Rosario Formation. One of these submarine-fan systems is best seen east of the highway at Km 92. At this stop, the sedimentary rocks record an upward transition from the slope (outer-mid fan) to the inner fan deposits. Here the section rests on a dark mudstone (Middle Mudstone Member) that is overlain by a diamictite unit 40 feet thick. This is overlain by 20 feet of inversely graded imbricated conglomerate that represents slump and channelized debris flows. Above the conglomerates are 100 feet of submarine turbidite flows composed of sandstone and mudstone that fill a small submarine canyon. The over all thinning and fining upward nature of this sequence indicates an inner-fan channel of a submarine fan.

CRETACEOUS ROCKS NEAR ENSENADA: Sedimentary rocks deposited in this area during the Late Cretaceous are derived from the uplifted Peninsular Ranges Batholith. These sedimentary rocks developed as a clastic wedge, which is thicker to the west, consisting of a fluvial and alluvial facies to the east and a deep marine facies to the west. During Turonian time, fluvial and alluvial sediments were deposited as fans; these deposits are known as the Redondo Formation. The overlying sedimentary rocks comprise the Rosario Formation which are Campanian-Maastrichtian in age (Late Cretaceous), and the contact with the underlying Redondo Fm. is unconformable. In Campanian time, sea level rose and deposited a transgressive sequence (beach to offshore deposits are represented). The shoreline was defined by steep bedrock cliffs with encrusting molluscan assemblages such as the rudistid-molluscan bivalve assemblage at Rincon de la Ballena (*See* 2:22:14, Punta Banda road). The sea level rose with the deposition of a thick shale-mudstone unit described by Kilmer (1963) in the type area until Late Campanian time. The sea level fell in Early Maastrichtian and formed huge deltaic sequences. This regression dumped large amounts of sediment on the steep narrow shelf and caused gravity-flow deposits that developed numerous submarine fan systems.

VIEW, FROM THE SOUTH, OF 2013 SLIDE REPAIR. *The highway actually sits on a causeway that does not rest against the prominent back cut.*

93   Landslide reconstruction – This is the site of major landslide that disrupted travel along the highway in late 2013 and nearly all of 2014. There is a place to turnout just to the south of the slide. Going south it sneaks up on you. If you want to view the slide from the top you can take the free road, just south of the Toll Plaza, to Quatro-quatro and have them take you to a viewpoint over looking the slide. Worth it!!!

95.2   This is the site of a major landslide that disrupted travel along the highway for a number of months. During that time the old road was used as a bypass. Workers have stabilized the landslide by dewatering and repaved the highway. It has proven to be relatively stable since then.

96   The topography along the highway from here to Kilometer 98 is very rugged with many irregular small hills. Each hill represents an individual landslide area on the main point of Punta San Miguel. The entire coastline in this area is actively sliding into the sea due to the undermining of the point by the ocean.

LANDSLIDE AREA OF PUNTA SAN MIGUEL
CLIFF IS AT TOP OF SLIDE AREA

98   This is the first view of the small fishing village of El Sauzal and the site of another massive highway failure that occurred in 1978. The slope failure on the hills to the left resulted in the loss of a number of residential homes that once overlooked the ocean. Look carefully to the left to see the remnants of some of these homes. These homes were built on landslide blocks. When the blocks failed (moved down toward the highway), the homes were destroyed. They regraded the slide area and placing rock at the base of the slide in the summer of 1997. Abandoned homes were still visible.

LANDSLIDE SURFACES offer attractive building sites to the unsuspecting developer because they are often flat with unobstructed views. The low density development on some of the stabilized blocks has not immediately created unstable ground. However, more development on presently stable landslides may create disastrous effects as shown by problems in the San Miguel area. When structures are built on landslides, there is an increase in

37

the ground water due to residential watering and waste disposal. When this additional water infiltrates into the landslide it acts to weaken, load, and may lubricate the slide mass so that it become unstable. The demand for beach-front property is increasing, and more of these stabilized blocks are becoming developed. More landslides, property loss, and possibly casualties are inevitable.

LANDSLIDE AT TOLL BOOTH IN SAN MIGUEL

98.5   Caseta de Cobro.

99.2   Junction of the Tijuana Libre Highway 1 and Highway 1D.

100   The highway between here and Ensenada follows the same Late Pleistocene Terrace that the road has been intermittently following since leaving Tijuana.

101   The hills to the left of the highway for the next several kms. are composed of upper Cretaceous marine sandstones that were deposited along an exposed rocky shoreline.

101.3   The road to the left leads 114 Kilometers to Tecate on Mexico Highway 3 (*See* log 15).

101.6   The fish canning industry once supported the village of El Sauzal. A harbor has been built here to shelter the fishing fleet. This was one of the enterprises of the late General Abelardo Rodriguez, a former Governor of Baja California and briefly the President of Mexico until the election of Lazaro Cardenas in 1933.

105   A quarry is located east of the highway. The rocks that formed the Late Cretaceous coastline are exposed in the quarry and adjacent ravines.

106.5   The Universidad Autonoma de Baja California (UABC) with its Escuela Superior de Ciencas Marinas is located to the right of the highway on Punta Moro. This very prestigious marine biology and Oceanography school obtained legal rights to the property where the school is built after the students forced the governor of Baja to give it to them. UABC is one of the most prestigious universities in Mexico.

The Centro de Investigacion Cientificia y Educacion Superior de Ensenada (CICESE) is the series of buildings on the hill to the left of the highway. This school also has an excellent reputation in Mexico and the United States.

107   This is the intersection of the route to Ensenada, by Calle 10a, and the coastal route to Ensenada. Geologically the coastal route is more interesting to follow. It provides beautiful views of Punta Banda and Islas de Todos Santos as you drive the last three kilometers along the coast into Ensenada.

109.7   This road cut exposes a massive andesitic breccia. The metavlocanic rocks in this area are metamorphosed volcanic flows, pyroclastic rocks and sedimentary rocks of volcanic derivation. The most common rock types were basic to intermediate tuffs and breccias, basaltic and andesitic flows, and volcanic graywackes. The age of the rocks in the Ensenada area is still in doubt. Similar rocks range in age from Late Jurassic to the north in San Diego County to Early Cretaceous south of Ensenada. Here they were quarrying the rocks to use to enlarge the harbor. During the quarry operation some of the cliff collapsed onto the highway, so a detour was built around it. The removal of part of this point has resulted in more wind in the harbor area.

109.8   **ENSENADA** has been a natural port for a long time and blossomed into a major port during the west coast shipping strike in 1974. The strike was a boom to this city and resulted in the enlargement of the harbor. With this enlargement came more business even after the end of the strike. The availability of large blocks of metavolcanic rock in Chapultepec hill has made the task of building the Muelle (breakwater) and enlarging the port easier and cheaper than importing rock from a more distant quarry.

110   This is the intersection of Ave. Gastelum and Calle Primera in Central Ensenada. Follow Mexico Highway 1D (Blvd. Gral Lázaro Cárdenas) southward along the Malecón on the right. The main tourist shopping district, the destination of many American tourists, is one block to the east. There are a number of good restaurants and hotels in Ensenada. My current favorite is Nico Saad's San Nicolas Inn and Casino with its Olympic pool. The rooms on the courtyard are the quietest in town.

LAS TRES CABEZAS: The Malecón monument commemorates three famous figures of Mexican history: **Benito Juárez**, a great president of Mexico and a full-blooded Indian; **Miguel Hidalgo**, the priest who issued the "Grito de Delores" that declared Mexican independence from Spain; and **Venustiano Carranza**, first president of the new Mexican Republic in 1917.

At the south end of the Malecón (Boulevard General Lázaro Cárdenas) the highway jogs several blocks to the left on Ave. General Agustín Sangínes. Then it turns right on Avenida Reforma (Highway 1).

If you have time you really need to visit **HUSSONGS CANTINA** *established in 1892. Two blocks up and one block to the left on Ave. Ruiz. Hussong's is reputedly the place where the Margarita was created in October, 1941 by bartender Don Carlos Orozco for Margarita Henkel, daughter of the German Ambassador to Mexico.*

Log 2 - Ensenada to San Quintín [196 kms = 122 miles]

The highway travels south on the Pleistocene terrace bordered by metavolcanic hills to the left and views ahead of the metavolcanic spine of Punta Banda and the Agua Blanca Fault Zone. It then drops into and crosses the fertile alluviated Maneadero Valley to climb back onto the terrace before turning inland up a valley through the metavolcanics to generally follow along the trace of the Agua Blanca Fault Zone. The highway crests a small pass where it crosses the Agua Blanca Fault Zone, and drops into and follows Santo Tomás Valley that is developed in the metavolcanics along the Santo Tomás and Agua Blanca fault zones. Santo Tomás and the grade south of town are in the Santo Tomás Fault Zone.

The highway follows a series of oak woodland valleys through more metavolcanic ridges with alluvial valleys and tonalite outcrops to enter the rolling mixed granitic hills and gentle alluviated valley areas of the San Vicente Plain. Beyond the San Vicente Plain, the highway descends a grade and turns seaward through metavolcanic hills to Colonet on the edge of the San Quintín Plain.

At Colonet, the highway turns south to parallel the ocean and follows the Pliocene-Pleistocene marine terrace developed on marine sedimentary rocks of the Rosario Formation and the Pliocene Cantil Costero Formation. The high hills to the east are part of the metavolcanic Alisitos Formation. They formed the shoreline during the development of the Pliocene-Pleistocene terrace. The highway will alternately climb onto the terrace and drop into alluviated valleys. South of Vicente Guerrero, the Baja highway drops onto the lower late Pleistocene terrace with views of Laguna Figueroa, San Quintín Bay, and the basalt cones of the San Quintín Volcanic Field. Near San Quintín, the higher marine sedimentary terrace forms a series of mesas backed by the high metavolcanic hills.

0   Start at the shopping center at the intersection of Gral. (General) Agustin Sangenes and Reforma = Highway 1. There may be no log markers for almost 16 kilometers until you reach the Maneadero Valley.

6   The highway heads south along Ave Reforma parallel to the coast, on the wide and well-developed 10-meter Late Pleistocene terrace. In front and to the right are views of the spine of Punta Banda and of the active coastal dune front that is partially stabilized by vegetation of the Chaparral region of the Californian Phytogeographic area.

The **metavolcanic hills** to the left consist of a well-preserved section of pyroclastic rhyolite and dacite in a large syncline that plunges to the west.

The stratigraphic section exposed here consists of 8,000 feet of lithic and crystal dacite tuffs, breccias, welded tuffs; welded lithic tuff and andesite; tuffaceous andesite breccia sandstone; and tuffaceous andesite breccia. A quartz diorite pluton limits the structure and stratigraphy on the north near the Cementos California plant, while a granodiorite pluton terminates the section on the south near the small village of Maneadero. No fossils have been found in this section, but its lithologic characteristics indicate that they are representatives of the early Cretaceous volcanic episodes that have been dated to the south of the Agua Blanca Fault.

12   Ensenada's military airport is for both military and commercial flights.

13   The road to the right leads to Estero Beach Resort on Playa Esteros.

16   The highway leaves the Terrace and descends through Pleistocene dune sands into the fertile Valle de Maneadero. As the highway crosses the Valley, it approaches the spine of Punta Banda and the Agua Blanca Fault that the highway will follow up a valley to the left.

18.5   The side road to the left leads to San Carlos Hot Springs and Resort (20 Kilometer) situated on a fault zone in the metavolcanic rocks.

Hot Springs: Hot springs are at least 10° C warmer than the mean air temperature. Hot springs do not normally require abnormal heat sources. The earth's temperature increases about 1.8° Centigrade per 100 meters of depth. As ground waters are heated to the temperature of the surrounding rock, the 10° warmer water need only come from about 550 m. below the

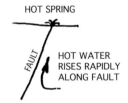

surface. Faults provide an avenue for the rapid rise of hot ground water to the surface which results in hot springs. Most of the hot springs in the western United States are along faults. Some such as Yellowstone and Lassen are near volcanic centers and are due to magma near the surface.

22   Turnoff to Punta Banda and La Bufadora (21 Kilometer). **The main highway veers left to leave the valley and climbs onto the coastal terrace. To continue south *See* Kilometer 23 below**.

SIDE TRIP TO LA BUFADORA AND PUNTA BANDA:

The side road turns right towards the coast and makes a number of right angle bends to avoid crossing cultivated fields. As the road approaches the hills, the trace of the Agua Blanca Fault Zone is at the break in slope between the steep hills and the gentle fan slopes.

7   This slough is a sag pond on the Agua Blanca Fault. The edge of

the slough is a fault scarp. The road crosses this scarp about 1/2 km to the southeast. Many fault features are visible along the base of Punta Banda Ridge *(See fault figure 2:24).*

7.5   At the highway curve, rushes *(Junco sp.)* are growing in on the side of the road. Numerous resident Red-winged Blackbirds are seen building nests here in the spring *(See 1:69.2).*

8   Castor Bean, Junco, Typha, composites, and Giant Reed are found in roadside ditches. Tamarix and olive trees form windbreaks. The area is an "estero" or salt marsh. Other distinctive plants in the area include Giant Reed *(Arundo donax* of the Toyon grass family), California pepper tree, Indian Tree Tobacco, Laurel Sumac, Rabbit Brush, and Datura.

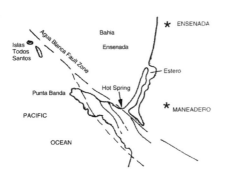

9   The sand spit separating the estero from Bahía Todos Santos can be seen ahead to the right. Numerous hot springs along the recent break of the Agua Blanca Fault lie at the edge of the estero a few hundred feet or so to the right of the road. The western-most hot spring is on the main beach and faces the ocean near Kilometer 11, almost opposite the northernmost house. It is possible to dig in the sand here at low tide and encounter hot water *(See 2:18.5).*

10.6   Turnoff to Punta Estero and the Baja Beach & Tennis Club on the sand spit. Pleistocene terrace materials are exposed in the roadcut. Pickleweed, Alanofria, and other halophytes grow in the sands of the estero.

The dunes are stabilized with Wild Buckwheat, Beavertail Cactus, Ice Plant, and annual sunflowers. Only halophytes grow in this region. Datilillo is planted around residences and gates to give more of a tropical flavor to the environment. Annual grasses take advantage of the highway edges.

The avifauna of the mudflat near the mainland in the upper part of the estero consists of Dowagers, Godwit, Cattle Egrets, Western Gulls, California Gulls, and Turkey Vultures. Many birds dig for invertebrates in the mudflats.

11   Road cuts of faulted and deformed dacite porphyry. They are part of a metavolcanic mass that is the backbone of Punta Banda. For the next three kilometers, the road traverses Pleistocene and upper Cretaceous rocks, faulted against or deposited on the older metavolcanic rocks on the north side of the point. The vegetation on the hills is typical of the chaparral flora

and consists of Cheesebush, Toyon, Wild Buckwheat, Black Mustard, Locoweed, Tamarisk, and Mission Cactus.

12.5   La Joya.

13.2   View to the left rear of the Agua Blanca Fault trace. The fault passes through the notch in the skyline and follows the base of the steep slope.

14.2   Cretaceous age rocks were first recognized and described in Baja California at this location. The road into the gully past the houses on the right leads to a small quay on the beach. The Cretaceous rocks are exposed in the sea cliffs near the quay and in road cuts for the next two Kms.

CORALLIOCHAMA ORCUTTI LOCALITY

Coralliochama orcutti: C.A. White (1885) described *Coralliochama* (aberrant clams) bearing rocks on Bahía Todos Santos. This genus is the most characteristic elements in the late Cretaceous near-shore faunas of Northwestern Baja California. Accumulations of specimens, some with shells attached and partially abraded, can be seen in these sea cliffs.

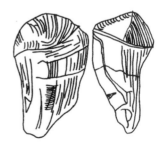

The upper Cretaceous beds at this locality are more highly deformed than at any other locality known in this region. This is due to their proximity to the Agua Blanca Fault that lies just offshore.

16.5　　Gray-green rocks of the Alisitos Formation exposed along the road.

17.8　　The road crests a grade and briefly passes along the spine of Punta Banda. The Parry Buckeye grows in arroyos and on the hillsides.

18　　The village of La Bufadora is below the road to the left. Several Pleistocene marine terraces are obvious along this segment of the point.

The vegetation on the terrace consists of Toyon, Wild Buckwheat, Sumac, Chamise, Agave, Jojoba, Broom Baccharis, and Indian Tree Tobacco.

20　　This area of Punta Banda is composed of massive highly-jointed metamorphosed andesite porphyry. The blowhole is developed along a large joint in the metavolcanic rocks. The contact between the andesite and the rounded boulder conglomerates in the Pleistocene terrace is exposed along the road.

21　　**La Bufadora:** Cafes and numerous small curio shops are located on Punta Banda. These establishments have developed to serve the visitors who come to view the activities of the world's largest blowhole.

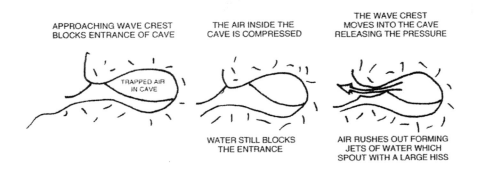

THE WAVE CREST MOVES INTO THE CAVE RELEASING THE PRESSURE

APPROACHING WAVE CREST BLOCKS ENTRANCE OF CAVE

THE AIR INSIDE THE CAVE IS COMPRESSED

TRAPPED AIR IN CAVE

WATER STILL BLOCKS THE ENTRANCE

AIR RUSHES OUT FORMING JETS OF WATER WHICH SPOUT WITH A LARGE HISS

THE LA BUFADORA BLOWHOLE is reputed to be the largest blowhole in the world. Waves approaching the shore of Bahía Papalote trap air in a rocky cave along a large fracture on the northern shore of the bay and block the cave mouth. As the wave crest moves into the cave, it compresses the cave air until the release of pressure along cracks in the roof forces a plume of water skyward. The steepness of the wave front and the amount of water blocking the cave entrance determine the strength and height of each plume of water. If there are no big swells, the blowhole may not put on a show.

LA BUFADORA BLOWHOLE

Another attraction can be experienced by divers just offshore in the form of hot springs that have developed along the Agua Blanca Fault. Some divers have hard-boiled eggs in the hot waters that rise out of the springs.

From La Bufadora, you return to the main highway at Km 22. You may turn right and continue to the south or return to Ensenada.

CONTINUATION OF MAIN LOG

24   There is a view to the north of the Miocene basalt mesas, Islas de Todos Santos, Bahía de Todos Santos, the point of Punta Banda, and the long spine of hills stretching southeast from Punta Banda along the Agua Blanca Fault.

Notice the long, straight nature of this spine, the flats along the spine, and the lines of vegetation. Near the middle of the spine and close to the highest point, there are a series of houses on a flat terrace. These houses sit directly on the Agua Blanca Fault. Note the alluvial fans coming down from the fault zone. On either side of the houses, there are offset streams that indicate right lateral movement. To the left is a notch in the hills through which the highway will cross into a valley occupied by the Agua Blanca Fault. The Diagram (from Borcherdt, 1975) depicts some of the fault features that you are viewing.

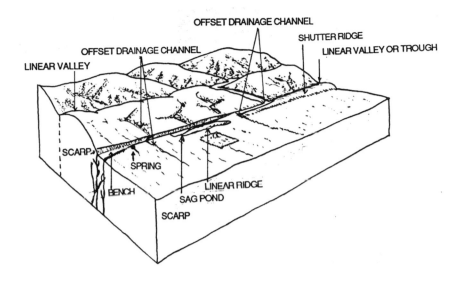

OFFSET DRAINAGE CHANNEL

SHUTTER RIDGE

OFFSET DRAINAGE CHANNEL

LINEAR VALLEY OR TROUGH

LINEAR VALLEY

SCARP

SPRING

BENCH

LINEAR RIDGE

SAG POND

SCARP

33   As the highway follows Cañon las Animas, it parallels the Agua Blanca Fault Zone that is located to the right. On the ridge above the highway are examples of shutter ridges, benches, vegetation lineaments, and offset streams that typically characterize a major fault zone.

Black-shouldered Kites (*Elanus caeruleus*) are commonly seen flying, hovering, and hunting along this stretch of highway.

Black-shouldered Kites are large falcon-shaped gray birds with black shoulders, a white underside, and a long white tail. They may be seen flying, soaring, or hovering like a kite along the highways as they hunt small reptiles, rodents, and large insects. Kites are uncommon residents within the coastal plains and valleys from San Quintín north, and occasionally south to Guerrero Negro.

Age of the metavolcanic strata: There are lower-upper Cretaceous fossils in the Alisitos formation along the peninsula south of the Agua Blanca Fault. North of the Agua Blanca Fault there are no lower-upper Cretaceous fossils in the metavolcanic strata. The next fossiliferous rocks are the Santiago Peak Volcanics in the San Diego area; they contain uppermost Jurassic fossils (Fife *et. al.*, 1967). Between the Agua Blanca Fault and San Diego County, there are no fossil or radiometric dates on the metavolcanic strata.

45.1   The highway passes through a deep road cut in the Agua Blanca Fault at the crest of a small pass. There is a sweeping view of the fault line valley of Valle Santo Tomás.

SANTO TOMAS FAULT RUNS ALONG AND FORMS THE STRAIGHT BASE OF THE HILLS ON THE FAR SIDE OF VALLE DE SANTO TOMAS.

46   The road to the right leads to La Bocana and Punta China (26 kms) where the upper Cretaceous Alisitos Formation was first defined by Santillan and Barrera, (1930) for the rocks near Rancho Alisitos. A cement quarry on Punta China has been developed in the limestones of the Alisitos Formation.

This graded road is the newer, higher road that has been cut into the side of the hills to avoid the frequently flooded and washed out older road of the valley bottom. There is a nice developed camping spot 5 Kms. down this road in the old Oaks and Sycamores at the old Misión Santo Tomás.

50.5   The highway crosses the broad fault line valley of Santo Tomás. The large riverbed of the meandering Río Santo Tomás contains only a small perennial stream except during periods of heavy precipitation when it may become a raging river. Some of the water originates as freshwater springs from subsurface groundwater that rises upward through the crushed rock produced by movement along the Agua Blanca Fault Zone.

This water source once supported the small Dominican Misión Santo Tomás and was the reason for the mission's location here. An expedition of Portola and Father Serra named this valley "Cañada de San Francisco Solana." In June 1794, Father Loriente founded a second mission, the Mission of Santo Tomás de Aquino, after abandoning the first mission 5 Kms. farther downstream. The ruins of the second mission are behind a house near the highway. The dirt road on the north edge of town goes down the valley to eventually reach Punta China and La Bocana over a rougher route.

Riparian Woodlands are vegetational communities that grow along streams and other drainage ways. Since the climatic regime over much of Baja is an arid one, the local occurrence of permanent standing or running water has a dramatic influence on the composition and quantity of vegetation. Large deciduous trees, shrubs, and herbs that only grow on the banks of such watercourses generally border localities with permanent standing or running water. Where river valleys are broad, the Riparian Woodland is correspondingly broad; at higher elevations where the water courses are narrow and the stream banks are steep, Riparian Woodlands may form a very narrow strip that may be only a few feet wide.

51   **SANTO TOMAS:** The small village of Santo Tomás dates back to mission times. It was the original site for the famous Santo Tomás Winery.

The El Palomar restaurant is directly in the Santo Tomás Fault Zone. This fault forms the south side of the valley and the Agua Blanca Fault Zone forms the north side. To the north and south of the valley, the steep hills are densely covered with Coastal Sage Scrub.

57   Numerous small fans that mark a trace of the Santo Tomás Fault have developed at the base of the hills to the right. Streams are cutting into the small fans due to relatively recent uplift of the hills along the fault line.

60.1   The highway climbs a grade where road cuts exhibit numerous sheer and crushed rock zones typical of the trace of a major fault zone.

AGUA BLANCA FAULT VALLEY

61   The turnoff near this kilometer marking provides a good view of the Agua Blanca Fault Zone that extends through the valley to the north. There is

a line of bushes and very small trees that cut across one of the points just above and to the right of the white ranch house on the opposite side of the valley. This vegetation line is replaced by the low notch to the west where the crushed rock of the fault zone was easily weathered and eroded.

The northern end of the Agua Blanca branches near here to form the north and south sides of Punta Banda. Between these two branches, a resistant sliver of the Alisitos Formation forms the spine and point of Punta Banda.

63   At the crest of the highway, there is a well-graded side road to the left that leads to a microwave tower on Cerro el Zacaton. This dirt road crosses the Agua Blanca Fault Zone.

73.3   The metamorphosed volcanic and sedimentary rocks of the Alisitos Formation are especially well exposed in the road cuts along this arroyo. Some of the holes in the rock are the impressions of the shells of gastropods that have been leached from the rock.

The streams and valleys in this area exhibit Riparian Woodland species of Oaks, Willows, and California Sycamores while the hillsides contain the typical Coastal Sage Scrub vegetation represented by Brittlebush, Laurel Sumac, Broom Baccharis, Chamise, Wild Buckwheat, and Dodder or "Witches Hair" growing on plants.

75.5   The resistant ridge that crosses the highway at this bend is formed by a series of light-gray limestones within the Alisitos Formation. These limestones are quarried at Punta China by a commercial concern, Cementos California. The limestones, which are low in magnesium and aluminum, outcrop in a belt that extends southeast from Punta China. The large reserves and their proximity to deep-water shipping facilities make it an important source of limestone for cement.

78 A paved road marked "Eréndira" leads south 6.6 kilometers down Cañon Guadalupe and Cañon San Isidro to Arroyo San Vicente, and then turns west to the coastal community of Eréndira and Bocana Eréndira.

SEA CLIFFS AND POCKET BEACHES: There are many beautiful coves and sandy pocket beaches with sea cliffs of Pleistocene terrace material overlying the Cretaceous Rosario Formation. They resemble the beautiful beaches at La Jolla, California. Generally, the sea cliffs in this area are formed in the Cretaceous Rosario or Alisitos formations. The pocket beaches form when the sea cliff is capped by softer marine terrace deposits. The fractures in the harder cliffs are enlarged into coves by the waves. This leaves a cove with a pocket beach that consists of coarse sand and gravel.

80.8   The highway crests over a ridge with a panoramic view of the San Vicente plain. Misión San Vicente Ferrer is located to the right before the bridge. This plain is often dry farmed for wheat and barley.

90   San Vicente originated in 1780 as a Dominican mission site, but it fell into obscurity until the 1940s. Rejuvenation of the community was the direct result of agricultural development in the area.

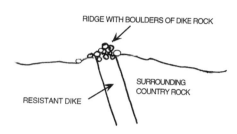

92   For the next kilometer, the low linear mounds of light-colored boulders represent the weathered remnants of resistant granitic dikes that filled shrinkage cracks near the top of the batholith. The country rock (tonalite) is less resistant and more easily weathered and eroded. This leaves the ridges of dike rock.

101   This flat plain is Llano Colorado (red plain). The view to the east is of the metavolcanic foothills. Llano Colorado supports some of the extensive grape vineyards and olive orchards in Baja.

107.5   The highway crosses an arroyo with dense Tamarisk that is replacing the Saltbush.

Saltbush is a halophyte that has modifications that enable it to grow in very salty dry soils. The seeds of the Saltbush contain chemical growth inhibitors that dissolve when exposed to sufficient amounts of water. These inhibitors ensure that the seeds will only germinate when sufficient quantities of water are available to allow the seedling to complete its life cycle.

Saltbush can grow in soils with a high salt content. The salt is absorbed into the plant through its roots. The leaves can be used as a flavoring for foods, and the parched ground seeds make a tasty coarse meal or fine flour.

109   Granitic dikes are exposed along the grade.

113   Descend through granodiorite and gabbro outcrops into Arroyo Seco where the highway turns to follow the arroyo toward the sea. Metavolcanic rocks of the Alisitos Formation are exposed along the upper Arroyo Seco.

126.4   Highway crosses Arroyo Seco and passes through Punta Colonet.

131.5   The highway dips into a small steep gorge with rocky cliffs of the metavolcanic Alisitos Formation. The low range of hills, west of the highway, are stabilized coastal sand dunes. The red clay soils in this area are derived from the iron-rich Alisitos Formation. The Llanos de San Quintín is drip irrigated and dry farmed. The highway begins to follow a flat 30-meter Pleistocene terrace often called Llanos de San Quintín.

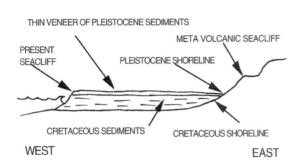

THIN VENEER OF PLEISTOCENE SEDIMENTS

META VOLCANIC SEACLIFF

PRESENT SEACLIFF

PLEISTOCENE SHORELINE

CRETACEOUS SEDIMENTS

CRETACEOUS SHORELINE

WEST

EAST

To the left are the foothills of the Alisitos Formation. The shoreline in the Cretaceous, when the terrace rocks were deposited, and during the late Pleistocene, when this terrace was cut, is roughly equivalent to the edge of the terrace and the base of the foothills.

140.9   The road to the left leads to the small village of San Telmo (10 Km), the Meling Ranch (50 Km), Mike's Sky Rancho, the pine groves of San Pedro Mártir National Park, and a National Observatory (87 Km).

OLD ROAD AND ITS CHOICES

Plutonic rocks of the Peninsular Ranges: Reconnaissance mapping in the Peninsular Ranges has resulted in the identification of 387 plutons over 1/2 mile in diameter that cover 28,000 square miles. These plutons are smaller on the western edge of the range while the axial portions of the range are occupied by a number of much larger bodies. Most of the relatively small plutons are circular in outline and many show concentric structures. Granite and gabbro form the smallest plutons, while most of the largest plutons are composed of tonalite. These plutons yield isotopic Uranium age dates of 95 to 119 million years before present (MYBP). K/Ar dates of 75 to 85 MYBP in the western edge and in the 60s on the eastern edge yield cooling dates or ages of metamorphism (Gastil *et.al.* 1975).

The rocks on the road to the Meling Ranch and the high San Pedro Mártir exhibit a cross section of the pre-batholithic and batholithic rocks of the Peninsular Ranges. The road begins in the metavolcanic and metasedimentary rocks of the Alisitos formation. This area is stratigraphically the lowest strata in the type area of the Alisitos Formation containing Albian fossils (Allison, 1974). Nearer to the ranch, these rocks are progressively more metamorphosed until they become schists and gneisses. At the ranch, a pluton of granitic rocks is exposed. As the road climbs to the observatory, it enters an extensive area of granitic rocks.

FLORA OF THE SIERRA SAN PEDRO MÁRTIR - The Sierra San Pedro Mártir averages 5,000-10,000 feet or more in elevation. Its western slopes are covered with coniferous forests that receive 20 inches of precipitation per year on the average. The dominant species of the coniferous canopy are Jeffrey Pine, Incense Cedar, Sugar Pine, Lodgepole Pine, and Colter Pine. At lower elevations, Quaking Aspen and Canyon Oak frequently occur.

At elevations below 3,000 feet, the western slopes of this range are covered by true Chaparral and Coastal Sage Scrub. The drier eastern slopes are sparsely covered with vegetation that is characteristic of the San Felipe Desert, a southern extension of the Colorado Desert of the United States. (The vegetation of the San Felipe Desert will be discussed at 14:108).

At the southern end of the Sierra San Pedro Mártir, the flora is in transition between the Chaparral vegetational area of the Californian Phytogeographic Region and the desert flora of the Desert Phytogeographic Region. As a result, the southern slopes are covered with a mixture of northern chaparral species and the desert vegetation of the Vizcaino Desert. This region of overlapping floras is called an ***ecotone*** and extends approximately 160 kms. from Valle Santo Tomás to El Rosario. This zone contains mixed Chaparral and desert flora which grade into each other at the respective ends.

141.3 The ecotone vegetation of this region consists of a mixture of the following Chaparral, Coastal Sage Scrub, and desert plant species:

Candelabra Cactus, Wild Buckwheat, Burrobrush, Burr Sage, Agave, Jumping Cholla, Brittlebush, Beavertail Cactus, and annual composites.

151.5 The highway climbs through more upper Cretaceous Rosario Formation outcrops. This area is an uplifted fault block (horst). Erosion of the area exposes the sedimentary rocks. Pliocene fossils have been collected from the Pliocene Cantil Costero Formation in the second and third quarry at Kilometer 155.

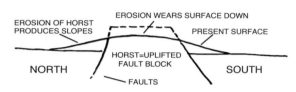

155 Some of the topography on the horst is due to stabilized coastal dunes. These dunes are much more obvious on the next rise to the south. At the crest of the rise, there is a view of the San Quintín Plain and the cinder cones of the San Quintín Volcanic field.

155 The highway begins a descent and passes through metavolcanic outcrops and then through Pliocene outcrops on the south side of the horst. Remnants of the road material quarries still produce Pliocene aged fossils.

Ten small cones of the **San Quintín volcanic field** are visible from the highway. Woodford (1928) was the first to describe the San Quintín volcanic field in great detail. This field is a series of olivine basalt cinder cones and flows of Pleistocene to Recent age. Locally, marine terraces are developed on the lava flows. This indicates that the eruptions predate the falling sea level and the 125,000-year-old Sagamonian Terrace.

157 Camalú.

166 Road cuts in this area are composed of dune sands. The rolling hills in this area are a stabilized coastal dune field that is extensively plowed and planted with various agricultural crops. This can lead to the destruction of the fragile dune environment and the development of "dust bowl" conditions if the plowing and planting is not done properly and at the correct time.

It has been noted that there is an increased amount of sand that blows across the road. This is due to the disruption of dune stabilization caused by farming practices (remember the U.S. Dust Bowl of the 1930s).

169.8 The highway crosses Arroyo Santa Domingo. This bridge was washed out in early 1978. It was rebuilt and a second bridge was added to accommodate future floodwaters.

The hills to the left of the highway are composed of the metavolcanic and metasedimentary strata of the Alisitos Formation.

170.2 Vicente Guerrero.

As the highway descends to traverse Llanos de San Quintín, the shallow Laguna Figueroa is visible behind the dunes along the coast of Bahía San Ramon. This area has been exploited for salt since mission times.

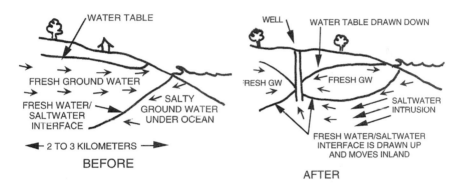

The San Quintín plain is extensively dry farmed and irrigated from groundwater wells. Over usage of the groundwater near the bay is resulting in salt-water intrusion; the salty groundwater under the bay pushes inland (intrudes) and displaces the fresh water as it is withdrawn.

183.1 Be sure to have Mexican auto insurance! John and Edwin were involved in an automobile accident near the store here. All premiums on Mexican auto insurance were returned when we slid into a truck that made an illegal left turn. **Be aware** that the Mexican insurance covers repairs that are done in their shops at their rates (usually on this side of the border). If you want to choose your shop, they will only give you their estimate at their cost of repair rates.

SAN QUINTÍN VOLCANIC FIELD

193   Lazaro Cardenas is the site of a military base.

196   The kilometer markers at Kilometer 196 change to 0.

MAMMALARIA CACTUS IN BLOOM

Log 3 - San Quintín to Bahía de Los Ángeles Junction. [285 kms = 177 miles

San Quintín to El Rosario - *South of San Quintín, the highway skirts a dune area and follows the low terrace along the base of the high mesa of Cretaceous marine sedimentary rocks overlain by Pliocene marine sedimentary rocks. Closer to El Rosario, these high mesas become much wider following a relatively straight Cretaceous shoreline. At Consuelo, the highway leaves the coastal terrace and climbs up a valley through Cretaceous hills to the high mesa surface with the metavolcanic hills in the far distance to the east. The highway briefly crosses a mesa before descending a steep canyon through the Rosario Formation to El Rosario.*

El Rosario to Cataviña - *This stretch travels almost directly east. From El Rosario, the highway turns inland up the Río Rosario on a river terrace, crosses the river, and passes through rolling hills of Cretaceous marine sedimentary rocks. The road climbs up on the rolling mesas with distant views of the resistant Paleocene mesas to the south and the high granitic San Pedro Mártir to the north. It then crosses the Cretaceous shoreline into the rolling metavolcanic foothills and descends a grade into a series of gentle valleys developed in the metavolcanic and granitic rocks. The highway then follows a broad valley with steep ridges of metavolcanic rocks on both sides of the valley. After El Progresso and a low pass in the metavolcanic rocks, the highway begins to traverse the San Agustin Plain on lakebed sediments. The steep metavolcanic hills to the left formed the shoreline of the lake. It then follows a flat to rolling surface developed on fluvial conglomerates with higher hills of granitic and metavolcanic rocks to the south. To the north is the extensive flat area of the San Agustin Plain broken by isolated low metavolcanic hills in the western part and volcanic mesas in the eastern part. Passing between several volcanic mesas, the highway enters the rolling hills of a picturesque bouldery tonalite area, finally descending through Arroyo Cataviña to Cataviña.*

Cataviña to L.A. Bay Turnoff - *The highway crosses a palm-studded arroyo to Rancho Santa Ynez and turns south through a rolling bouldery tonalite. The volcanic mesas become more numerous until, at Jaraguay, the highway climbs to cross an extensive undulating basalt plateau dotted with cinder cones. The bouldery hill of Pedregoso rises through this plateau. The highway then skirts the plateau in mixed granitic and metamorphic rocks and drops into the flat alluvial valley and lakebed of Laguna Chapala with dunes at the south edge of the lakebed. After a brief climb into a low pass in the tonalite hills, the highway crosses the Peninsular Divide with views of a wide valley and of the Sierra de la Asamblea. It then drops into the alluviated valley through tonalite and then Miocene fluvial sedimentary rocks to the granodiorite hill of Cerrito Blanco. After again passing over a very gentle Peninsular Divide, in alluvium, the highway begins to descend the Arroyo Leon drainage through the Miocene fluvial sedimentary rocks to the junction with the highway to Los Ángeles Bay.*

0   The kilometer markers at Kilometer 196 change to 0.

1   This well marked turnoff is the best road to take to the Old Mill (Molino Viejo) and Ernesto's Motel on Bahía San Quintín. Around 1885, an American company tried to settle the area. They built a flour mill and pier and tried dry farming the area for wheat. Today modern drip methods are providing for a thriving agricultural industry catering to U.S. markets.

5   The highway swings to the left to cross the arroyo over a high bridge near the hills. The hills to the left are part of the Rosario Formation and are overlain locally by the Pliocene Cantil Costero Formation.

Look and listen closely for Western Meadowlarks (*Sturnella neglecta*). The **Western Meadowlark** breeds in summer in the northeastern part of Baja and winters throughout the peninsula where they are commonly seen and heard in fields and on fences. The Western Meadowlark is a heavy-bodied, medium-sized bird. The most distinctive identifying field characteristic is a black "V" on its bright yellow breast. The voice of the Western Meadowlark is a rich loud and bubbling whistled song that suddenly interrupts the silence of Baja's desert landscape.

5.5   WASHOUTS: In the past, the main road went straight ahead through the wash. After repeated washouts, most of the traffic followed an easier route that crossed the wash at the base of the hills near Consuelo. When the highway was finally paved in the early 1970s, it once again went directly across the wash where it was again repeatedly washed out and finally permanently located nearer the hills.

8   The highway approaches the Rosario sea cliffs where it roughly parallels an ancient shoreline. It is built just above what would have been the old shoreline when the Llanos de San Quintín was cut as a marine terrace.

9   The highway crosses the large Arroyo San Simon by a high well-built bridge. If you want to see the red bed and mudstone outcrops of the Rosario Formation, park just before the bridge and walk about 100 m. into the road material quarry.

CHANGING SHORELINES: South of Colonet, the metavolcanic hills at the edge of the terrace formed both the Cretaceous and Pleistocene shorelines. South of San Quintín, the edge of the low terrace represents only the Pleistocene shoreline. The Cretaceous shoreline has undergone a greater amount of uplift and is on the level of the upper terrace. A climb onto this terrace will present a view of the Cretaceous shoreline that is fringed by the same metavolcanic hills that are seen to the north.

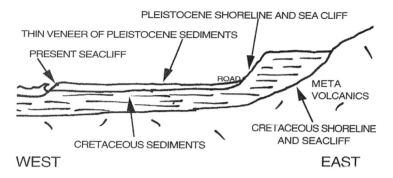

PLEISTOCENE SHORELINE AND SEA CLIFF

THIN VENEER OF PLEISTOCENE SEDIMENTS

PRESENT SEACLIFF

ROAD

META
VOLCANICS

CRETACEOUS SHORELINE
AND SEACLIFF

CRETACEOUS SEDIMENTS

WEST   EAST

11   The Tamarisk-lined side road to the right (just past the Pemex Station) leads to Hotel Mission de San Quintín and Cielito Lindo Motel and RV Park. Juanita, who ran Cielito Lindo, was one of the best hostesses (now retired) that the authors have found in Baja. Some say that they have never had a better meal than her rock crab claw plate. For many years, this hotel was the site of an American movie colony retreat that was owned by Mark Armistead, a cinematographer and inventor of the instant replay. The more notable frequent guests were Jimmy Stewart, Henry Fonda, and John Wayne who actually worked on one of the rooms.

TAMARASK-LINED LANE

TAMARISK TREES (SALT CEDAR): The side road which leads to Cielito Lindo is a very picturesque lane that is arched by two parallel wind rows of old Tamarisk trees that were planted to form fast-growing wind breaks for the agricultural areas of this region. A ride down this road at night is a little eerie.

59

Tamarisk Trees (*Tamarix pentendra*) were introduced into North America in the 1800s for use as decorative plants and fast-growing windbreaks. A single plant may produce 600,000 seeds per year. Its leaves secrete salt, an adaptation to reduce the tree's salt allowing it to live in saline soils. Surface soils under the tree accumulate salt that inhibits the germination and establishment of non-salt-tolerant natives. The quantities of duff (litter) produced by the Tamarisk encourages fire. They resprout from their own roots following a fire. Soils inhabited by Tamarisk soon become arid, as each tree transpires hundreds of gallons of water daily from the soil. These factors aid in the spread of the Tamarisk and reduce the numbers of native species.

BEACH DUNES: The beach dunes in this area are migrating to the south along the back beach. They are 3-6 feet high and continually migrate. During high tides, the dunes are often wave-leveled, forming a flat surface. As the beach dries, the sand again forms migrating dunes.

SAN QUINTÍN BEACH

WARM SANDY BEACHES AND HIGH TIDES: The warm sandy beaches along Bahía Santa María are some of the finest in Baja. They are wide and flat and are covered with sand dollars and Pismo clam shells. At low tide, the live sand dollars form interesting and unusual patterns as they move their spines to burrow into the sand. If you are careful, you can cross the dunes at the Mission de San Quintín and drive on the beach for several kilometers in each direction.

SAND DOLLAR TRAIL

If you drive to the west along the sand spit that forms the southern edge of San Quintín Bay there is a view of volcan Media Luna (half-moon), one of the basaltic cinder cones of the San Quintín Volcanic Field.

MEDIA LUNA CINDER CONE

Hotel Mission San Quintín used to have a swimming pool and a sea wall with a downstairs bar and restaurant. In 1976 high tides and a tropical storm put

61

three feet of water into the lower story of the hotel and filled the pool with sand. The sea wall is gone and the dunes are encroaching upon the area.

The brackish water marsh at the Hotel Mission has expanded over the years. Coots, Cormorants, California Gulls, Western Gulls, ducks, and several species of Grebes are commonly seen in or near the marsh.

The marsh vegetation consists of salt-tolerant halophytic plants such as Sand Verbena and the succulent Pickleweed (*Salicornia pacifica*). This salty plant is edible in salads and can be boiled and used as a pickling mix.

19   The hills of the Pleistocene shoreline converge with the road as the low terrace narrows and disappears under a dune field.

20   The highway passes Rancho Las Parritas and climbs onto the Socorro Sand Dunes. The source of the sand is the beaches near Cielito Lindo and Bahía San Quintín (see 3:10). When the migrating dunes reach the curve in the coastline to the southeast, they are blown inland.

21   At the crest of the first hill there is a sweeping view to the rear of the San Quintín volcanic field and the flat Llanos de San Quintín.

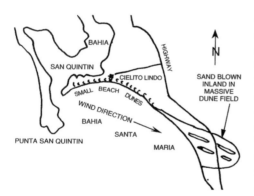

SAND DUNES IN BAJA: In the deserts of Baja, wind often heaps sand particles into the mounds and ridges of the sand dunes. Dunes are found wherever there is abundant sand and where the wind and the wind direction is fairly constant. They result from local interruptions in the general wind-flow patterns. Once they are formed, dunes tend to migrate slowly with the prevailing wind.

Sand is transported leeward where it gathers at the dune foot and constantly encroaches on new territory. Note that not all dunes travel. Where winds seasonally reverse or are multidirectional, they tend to remain stationary. Vegetation often helps to stabilize smaller dunes, particularly those that have developed around an embedded plant.

Although a sand dune looks like an inhospitable dry environment, it supports trees such as Mesquite, Tamarisk, and Catclaw Acacia; bushes such as Jojoba, Four-winged Saltbush, Burro Weed, Creosote Bush, Desert Buckwheat, Golden Bush, and Mormon Tea; and Cacti and grasses such as Rice grass and Big Galleta. Annuals carpet the dunes during rainy parts of the year. Some of the most commonly seen annuals on dunes in Baja are the

Sand Verbena, Evening Primrose, and composites. These bushes, grasses, and annuals enable the dunes to remain where they are by stabilizing the sand with their branches and roots. Although dunes may look dry, they are not. All precipitation that falls soaks in and very little runs off. Though the surface dries quickly, giving the impression that the interior is moistureless, water remains in the dunes long after the regions surrounding them are dry. This provides moisture for plant growth.

The dune vegetation provides shelter, food, and protection for numerous animals including insects, centipedes, scorpions, birds, lizards, rodents, rabbits, gophers, and ground squirrels.

THE HIGHWAY EDGE EFFECT: In the disturbed soils along the roadside, there are several mesic opportunistic plant species. The most obvious are yellow sunflowers, gray Atriplex, Burro Weed and small herbaceous annuals. These opportunists are seen growing not far from the highway.

In the desert regions of Baja, the dominant vegetation consists of plant species adapted to long hot dry summers. Dry adapted desert species that require little water are known as xeric (meaning dry) plants. In a region as rigorous as the Californian and Desert Regions of Baja, microclimates become particularly important. Any fortuitous or fortunate circumstance that provide habitats a little less xeric, where moisture remains somewhat longer than in more open environments, favors the growth of opportunistic vegetation. Highway pavement edges are classic examples. Taller, greener, and more mesic plants flourish along the road edges because the additional moisture that is needed to support these mesic species that are not adapted to the dry desert is available as runoff along the edge of the highway.

The plants commonly seen on the sand dunes are Ephedra, grasses, composites, Bursage, Burroweed, Jojoba, Ice Plant, and species of cacti. The dune vegetation is usually less than 3 feet tall and fairly dense.

23   View across the dune field of rows of stabilized longitudinal dunes with unstabilized dunes closer to the beach and the San Quintín volcanic field.

24.5 Road to Rancho El Socorro.

25   The highway continues south on a Pleistocene marine terrace cut into the Cretaceous Rosario Formation. Due to uplift, the Cretaceous shoreline is now at a higher elevation and several kilometers inland. It will be crossed near Km 83 south of El Rosario. The hills to the left consist of Rosario Formation mudstones in the lower part overlain by the sediments of the Paleocene Sepultura Formation. There are quite a few different kinds of plants growing here. Purple Bush is the dominant plant of the flats. There are specimens of Desert Thorn, Velvet Cactus, Agave, Cholla, and Bursage.

Tamarisk is the plant of the wash bottoms, and red-colored Ice plant and young Brittlebush grow in the disturbed soils at the pavement's edge.

30   The small hill to the right on the sea cliff is an old stabilized dune. It is being reactivated as the vegetation is being removed by the erosive action of the sea. *Other dunes can be seen near kilometer 41.* The sea is cutting into the base of the sand dune which demonstrates that there has been regression of the sea cliff since the formation of the dune.

36.2   Arroyo Hondo exposes Pleistocene Terrace material.

CANTIL COSTERO FORMATION IN CLIFFS

41 The large gray blocks on the slope are fossiliferous conglomeratic limestone of the Cantil Costero Formation named for these exposures of marine Pliocene rocks that cap the mesas along the highway. There are fossils in many places slightly above the limestone that include specimens of the large barnacle *Balinas tintinabulum*, which may be 6 inches tall. Distinct wave-cut terraces seen in this area were cut over a period of time by the sea. The highest terrace is developed on the Cantil Costero Formation (Late Pliocene). The Pliocene shoreline occupied nearly the same position as the present shoreline, but has been uplifted approximately 300 feet since it was formed. The cutting of the marine terrace into the Cantil Costero Formation during the Pleistocene has exposed the old shoreline. The lower highway terrace corresponds with the Sagamonian high sea level stage 125,000 years ago. Along the shore, a pair of dunes have been cut by the sea. The southern dune is still partially stabilized.

VELVET CACTUS   **GROUND COVERING SUCCULENT**

42.8 The highway turns inland at El Consuelo and climbs through the subdued hills of the Rosario Formation to the top of the mesa. This is the last view of the Pacific Ocean for several hundred kilometers.

The vegetation becomes denser. The plant species commonly seen along this stretch of the highway are Burrobrush, Pitaya Agria, Agave, Garambullo, Beavertail Cactus, Brittlebush, Nipple cactus, *Ephedra*, Velvet Cactus, Hedgehog Cactus, and occasionally Candelabra, Maguay, Cliff Spurge, and Milkweed. Locoweed (*Astragalus sp.*) is the plant that grows abundantly along the pavement's edge.

Astragalus is commonly known as locoweed because it contains an accumulative toxin, Selenium, that is poisonous to practically all livestock. They eat the toxin and go "loco" prior to dying from selenium poisoning. Although it is poisonous to cattle, early Baja Indians chewed the shoots to cure soar throats, used locoweed poultices to reduce swelling, and boiled the roots which produced a decoction that was used to wash granulated eyelids and ease the pain of toothaches. "*Astragalus*" comes from Greek and means anklebone, an early name for leguminous plants. The plants name may have referred to their low prostrate growth habit that placed them at ankle height.

Many species of Locoweed grow in the Californian Phytogeographic Region of Baja. They are distinguished by the length of floral parts and the size of the wooly seedpods. *Astragalus*, members of the pea family (F. Leguminosae), exhibit a variety of colored flower spikes (purple, pale yellow, and white) and occur in many different types of plant communities. Their leaves are alternately arranged along the narrow stems and are unequally pinnately compound (the leaf veins give the leaf a feather-like appearance). The dry pods rattle.

LOCOWEED

49.5   The highway climbs a grade to the top of a mesa through mudstone exposures of the Rosario Formation.

51.5   At the top of Mesa las Cuevas the vista to the northeast consists of mesas composed of Paleocene rocks overlying the Rosario Fm.

53.5   The highway descends a grade into Cañada el Rosario through the sandstones, mudstones, and conglomerates of the Rosario Group.

Rosario Group: Kilmer (1963, 1965) mapped and named four units of Late Cretaceous age: the Rosario Formation is the upper marine unit, El Gallo Formation is the underlying nonmarine unit, Punta Baja Formation is a lower marine unit and La Bocana Roja Formation is a the basal nonmarine unit. All four formations were part of Beal's original Rosario Formation. The thickness of the Rosario Formation near El Rosario is 10,000 feet.

Collecting in the vicinity of El Rosario (has been) a continuing project since 1965. Morris (1971) has described vertebrate fossils from the El Gallo Formation and indicates that "The fauna is small due to difficulties in collecting, scarcity of specimens, and the almost unique sedimentary framework. Some are 17 m. long and more aquatic than terrestrial. A large carnivorous dinosaur has been recognized as the result of the recovery of isolated teeth and cranial material. It is morphologically near *Gorgosaurus* (The tyrannosaur *Albertosaurus S.F. tyrannosaurinae* is more widely recognized by its synonym *Gorgosaurus*.).

"TERRIBLE LIZARDS" DINOSAURS: Dinosaurs evolved from reptilian ancestors in the early Triassic about 235 million years ago and became extinct at the end of the Mesozoic during the Maastrichtian stage of the late Cretaceous 65 million years ago. The earliest dinosaurs were members of a group of early crocodile-like thecodonts known as archosaurs. Eventually, dinosaurs populated all of the large landmasses except Antarctica.

Paleontologists have delineated two orders of dinosaurs - Saurischia (lizard hipped) and Ornithischia (bird hipped). It has been determined that if a strictly phylogenetic classification were used, it would be correct to say that modern birds are descended from the Saurischians and are living dinosaurs.

The American West presents the richest and most varied collection of dinosaur fossils anywhere in the world. The United States has become the center for the study of dinosaurs, and most of the field research and the interpretation of the bones takes place in the U.S. Near El Rosario, Baja California, several dinosaur bones and tooth fragments have been found and assigned to the Family Hadrosauridae. A *T. rex* tooth has been found in the area.

Do not collect Baja's fossil materials, since they are needed to illuminate Baja's past. It is against Mexican federal laws to collect the peninsula's resources!

MAMMALS: Although Cretaceous mammalian fossils are rare, they have been found in Baja. Specimens consist of isolated teeth, but several jaws have been collected and are related to forms found in Cretaceous deposits that outcrop along the eastern side of the Rocky Mountains.

REPTILES: A crocodilian which is close in ancestry to the lineage that leads to modern alligators has been uncovered.

BIRDS: A very significant avian fossil has been collected. This specimen represents the only terrestrial bird recovered from Mesozoic strata other than the famous *Archaeopteryx* from the Solenhofen Limestone of Germany. Studies indicates that it will serve as a phylogenetic link between the Jurassic *Archaeopteryx* and modern terrestrial aves."

54.5 The canyon on your left contains exposures of the Rosario Formation. In the right fork of this canyon, *Ammonites* have been found high on the slopes and ridges.

57.5 El Rosario is a small fishing and agricultural community. The road to the right at the corner leads to the ocean at the mouth of the Río El Rosario. A branch crosses the arroyo to the old mission ruins and ends at Punta Baja.

58   The highway turns left to follow a river terrace in Arroyo El Rosario, to Kilometer 61. Rocks of the Rosario Formation are exposed on both sides of the arroyo.

Across the Arroyo to the south there is a series of flat "benches." These benches are river terraces formed by the downward cutting action of Río del Rosario. The land in this area is gradually rising relative to sea level. This enables the streams to cut deeper into their beds and leave behind parts of their former channel. The terrace the highway follows is one of these terraces.

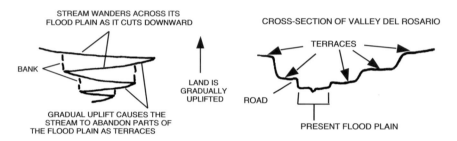

To the east of El Rosario along the left side of the highway, farmers often cover the hill slopes with chili peppers and allow them to dry in the sun before marketing them.

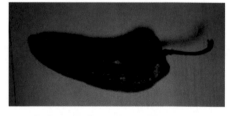

"CHILIES PEQUEÑOS" = SPICY, HOT CHILI PEPPERS!: Capsicum (chili) peppers, members of the potato family (F. Solanaceae), are used as a spice to flavor foods in both the New and Old World. There is a wide range of varieties. Peppers are New World natives that originated in Central Mexico and South America. They have been eaten for at least 7,000 years and have probably been cultivated for almost as long. Capsicum peppers are known from pre-historic burial sites in Peru, and they were widely cultivated before the arrival of Columbus in the 1490s.

The flavor and aroma of spices are due to essential oils. These oils of varied composition tend to have relatively small molecules. This makes them volatile at low temperatures. These oils belong to a group of hydrocarbons known as terpenes. The terpene of capsicum peppers is a volatile phenolic compound called capsaicin ($C_{18}H_{23}NO_3$), a substance so powerfully affecting human taste buds that even a dilution of 1:1,000,000 is detectable.

The greatest concentration of capsaicin is found in the pepper's placenta, the tissue that joins the seeds of the capsicum pepper to the fruit wall. Seeds of capsaicin peppers also have a lot of capsaicin. The fruit wall (ovary) has the least capsaicin. Because its presence or absence is due to variation in a single gene, some peppers lack capsaicin.

South of El Rosario, the vegetation changes dramatically with the appearance of the first endemic Cirios (*Idria columnaris*).

PENINSULAR ENDEMISM: Peninsulas such as Baja often have endemic (indigenous or native) plants and animals that are confined to the peninsula and are not found elsewhere in the world. Endemism is common due to the semi-isolation Baja California has experienced in past geological times. As a result many spiders, plants, and reptiles are peculiar to the Baja Peninsula. Endemism is especially evident among the cacti. Over 110 species of cactus have been reported in Baja; 80 of these are found nowhere else in the world. Other notable Baja endemics include the Cardon cactus and Elephant Tree. Endemism contributes to the picturesque uniqueness of Baja's vegetation.

DESERT ANNUALS: After rainy periods, whose times and amounts vary from desert to desert, the landscape is covered briefly by colorful "blooms" of annual ephemeral desert wildflowers. Their variously colored flowers (primarily white and yellow) stand out vividly against the earth tones of Baja's desert scape. Wildflower blooms are most often seen in the early spring in northern Baja and in the summer in the southern part of the peninsula. Annuals cannot withstand drought and may lie dormant for decades until the right conditions occur for growth. Then, the desert "blooms" with colorful floral displays. It is the function of the seeds to be the source for the next generation. These plants, which are so characteristic of the desert, evade the harshest desert conditions of Baja.

64.5 A high bridge crosses Arroyo El Rosario. The hurricane of 1967 did extensive damage to the El Rosario area. In early 1979 before the bridge was built, the paved highway was completely washed out and some travelers waited for four days to cross. Vehicles either forded or were towed across the river by a Caterpillar tractor.

65 **EL CASTILLO:** An interesting castle-like formation complete with battlements (dubbed "El Castillo") can be seen upstream to the left as the highway crosses the bridge. Movement on a fault parallel to the arroyo has resulted in the veneering of the canyon side with resistant flat-lying conglomerates that form the castle. The yellow-brown band high on the hills behind the castle is the boundary between the Cretaceous and Paleocene rocks in this area.

69

EL CASTILLO

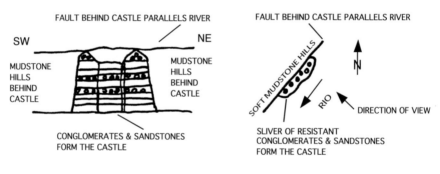

FAULT BEHIND CASTLE PARALLELS RIVER

SW   NE

MUDSTONE HILLS BEHIND CASTLE

MUDSTONE HILLS BEHIND CASTLE

CONGLOMERATES & SANDSTONES FORM THE CASTLE

FAULT BEHIND CASTLE PARALLELS RIVER

SOFT MUDSTONE HILLS

RIO

N

DIRECTION OF VIEW

SLIVER OF RESISTANT CONGLOMERATES & SANDSTONES FORM THE CASTLE

The road up the north side of the arroyo leads to the small Rancho Provenir. The majority of the hills along Arroyo Rosario consist of the mudstones and sandstones of the Rosario Formation. For the next 20 kilometers, the highway climbs through subdued exposures of the Rosario Formation.

71   The road cuts located on either side of this kilometer provide a good opportunity to stop to inspect excellent exposures of the mudstones of the Rosario Formation. The sandstones and mudstones of the Rosario Formation are exposed in the road cuts along this grade. The highway ascends a canyon and passes the first occurrences of specimens of Cirios (*Idria columnaris*) on the left.

73.5 A "grove" of cirios is growing on the hills to the left of the highway. Parry Buckeye are growing on the northern slopes along the highway.

BAJA'S WEIRD CIRIOS: The cirio trees of Baja are claimed by some as one of the endemics of the peninsula. However, a small colony of cirios are also found on the mainland of Mexico south of La Libertad. It is one of the most distinctive and interesting plant species in Baja.

This tall, candle-like plant derives its name "Cirio" from the Spanish word for wax taper (candle). It has been called a "boojum" because of its similarity to the fanciful creature described in Lewis Carroll's *The Hunting of the Snark*. Cirios have trunks that taper to a point approximately 60 feet above the ground in the largest specimens. However, due to the arid conditions of the peninsula, specimens rarely reach their maximum height. The plant looks somewhat like an upside-down carrot, with root like limbs that curve up and down in a grotesque fashion.

A story of doubtful authenticity has been told by some Baja travelers in an attempt to explain the strange shapes of the whip-like, grotesquely curving limbs. "This desert is very dry, so dry that when there is a fog (Neblina), there is more moisture in the air than in the ground. In order to survive, the Cirio trees turn upside-down so that their 'roots' are in the air in order to reach the fog. Since the fog layer is thin, the 'roots'that grow above the fog bend toward the ground to get back in the moisture-laden fog. As they grow toward the earth and grow below the fog, they turn upward away from the ground."

74   Good exposures of Rosario Formation conglomerates in road cuts.

74.5   The highway crosses a large arroyo with a good view of the varieties of vegetation which are characteristic of the Vizcaino Desert Region.

DESERT PHYTOGEOGRAPHIC REGION: The Chaparral flora of the Californian Phytogeographic Region declines near Valle Santo Tomás. In the 160 kilometers between Valle Santo Tomás and El Rosario, a transitional (ecotonal) flora consists of a mixture of Chaparral and desert species. Extending to the south from El Rosario east to the Gulf and to the south to La Paz is the drier Desert Region.

In contrast to the 10 to 20 inches of annual precipitation received by the more northerly Californian Phytogeographic Region, almost no rain falls in the Desert Region for two or more years. This region is truly a desert and is characterized by low humidity, widely fluctuating high ambient air temperatures, high surface and soil temperature, low organic content of the

71

soil, strong winds, high mineral salts content, and scarcity of water.

As the highway travels to the south through the Desert Region, it passes through three of the four subregions that are recognized and characterized by plant species. These three subregions are the Vizcaino Desert, the Gulf Coast Desert, and the Magdalena Plains Desert.

The highway between the Pacific coastal towns of El Rosario and Santa Rosalía on the Gulf coast passes through both the heart of the Vizcaino Desert and a portion of the northernmost extension of the Gulf Coast Desert. The extremely dry **Vizcaino Desert** covers the vast plain in west-central Baja California and extends from El Rosario to the palm oasis of San Ignacio. The floral dominants of the Vizcaino Desert Region are Cardon, Cirio, Century plants, Maguay, Datilillo, Agave, Yucca, Burbrush, and Ball Moss.

VEGETATION OF THE VIZCAINO DESERT

The **Cirio** or Boojum is the tallest plant of the visible flora. It looks like the taproot of a plant that has been turned upside down. *See* 3:73.5 to review the details about this strange plant.

Nests of hawks and Osprey are often seen in the upper branches of giant columnar cactus. Hawks, buzzards, and ravens use the Cardon as hunting, resting, sunning, or sleeping perches.

| CARDON | PITAYA AGRIA |

CARDÓN - The Cardon is a true Baja endemic. It is found from El Rosario to the tip of the peninsula and is the most widely spread of the peninsula's larger vascular plants. Since cactus lack leaves, photosynthesis takes place in the modified epidermal cells (chlorenchyma) of the trunk. The trunk is a true cladophyll (stem acting like a leaf). Stands of Cardon are called

Cardonals. Cardon is a cactus similar to the Saguaro cactus of the Sonora Desert of mainland Mexico and the deserts of the southwestern U.S.

PITAYA AGRIA is a dark green-gray stemmed cactus that grows in dense thickets from Ensenada to the Cape Region. The spines are reddish-gray with darker tips. The fleshy red fruit is edible.

GARAMBULLO   **ORGAN PIPE CACTUS with CARDON**

GARAMBULLO "Old Man" Cactus is an erect colonial cactus "tree" that grows to 12 feet tall and usually branches near the base. The tips of the star-shaped stems are densely covered with many long gray coarse hairlike spines that give the cactus a whiskered "old man" look. The fleshy red fruit is edible, but is not nearly as good as the fruit of Pitaya Agria. It occurs from south of Río Del Rosario to the Cape Region.

ORGAN PIPE CACTUS is a many-branched erect cactus without a main trunk. The Organ Pipe Cactus, which branches nearer the ground, lacks the main trunk characteristic of the Cardon. It looks like a set of organ pipes. The fleshy watermelon flavored red fruits are used as a food source in the late summer and fall.

BALL MOSS or Gallitos, a member of the pineapple family, is an herbaceous epiphyte that commonly grows on other plants but does not damage it or derive nutrition from it. In this region it is a common commensalistic epiphytic plant that grows on Cirios, cacti, shrubs and even telephone wires (*See* 16:16.5). Ball Moss is disseminated from plant to plant by wind-borne sticky seeds or by birds' feet or bills.

BALL MOSS OR GALLITOS   **RAMALINA LICHEN**

RAMALINA is a lichen. Lichens are a partnership formed by a combination of two plants consisting of microscopic photosynthetic green algal cells that live inside the cells of a non-photosynthetic fungus (mushroom). The relationship formed is known as a mutualistic symbiosis. The fungus provides water and protection for the algal cells. In return it utilizes the sugars, which are produced by the photosynthetic algae, for its source of energy. In this area of Baja lichens grow on Cirios or encrust boulders. Different species are visible as vivid red, silver-gray, and even black splashes of color.

AGAVE or Century Plants are members of the Amaryllis family. In Baja, four species of the genus *Agave* are commonly referred to as Agave or century plants. It takes determination and a brief study to tell the species apart.

In the Yucatan Peninsula, cultivated species of *Agave*, also known as "green gold," were once extensively cultivated for the fiber henequen (sisal). Botanically, the *Agave* fibers are long sclerenchyma, modified parenchyma fibers much like the "strings" found in celery. The immature flowering stalks and basal crowns are roasted. The seeds are ground into a meal.

80   Mesa la Sepultura "tomb" is the flat feature that is straight ahead.

80.2 This side route leads across Mesa San Carlos into Arroyo San Fernando to Abelardo Rodriguez (19 miles) and San Vicente. It ends on the coast at the base of Mesa San Carlos at Puerto San Carlos (36 miles). It is a long dead-end route, but the solitude and surfing make the trip worthwhile.

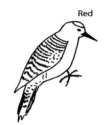

Red

The **Gila Woodpecker** is a medium-sized, red-capped woodpecker with a black and white barred back. It can be seen perching or flitting among willows and cottonwoods in riparian habitats, in desert cardonals, and stands of mesquite. This is the only woodpecker with a plain gray-brown belly, head, and neck and white wing patches that show during flight. They frequently bore nest holes in Cardons, mesquites, and cottonwoods. Gilas feed on insects and the fruits of mistletoe and cacti.

As the highway drops into the wash, Cardon, Rabbitbush, Burbush, and Wash Woodland-type vegetation predominate the landscape. On the left, just after crossing the wash, the typical Vizcaino Desert flora is represented by Cardon, Cirio, Pitaya Agria, Ocotillo, Garambullo, Agave, Organ Pipe, Cheesebush, Burro Bush, Broom Baccharis, Mesquite, and epiphytic ball moss growing on the Cirios. If you take a walk through the vegetation, you may see and hear the flitting red-capped Gila Woodpecker (*Melanerpes uropygialis*).

The highway curves to the northeast and drops through prominent exposures of conglomerate into the main wash of Cañada El Aguajito near the old Rancho of El Aguajito. Conglomerates that contain lenticular sandstones of the Rosario Formation are exposed in many places in this wash.

83.5 The contact between the Rosario and the Alisitos Formations is two-thirds of the way up the grade, southeast of the wash. There is an obvious change in the vegetation from the mudstones and conglomerates of the Rosario Formation to the reddish-brown soils of the Alisitos Formation. This change is due to differences in soil chemistries, soil moisture, soil salinity, and other soil edaphic factors.

85   At the top of the grade is a sweeping view. A series of concordant summits of the mesa tops can be seen to the northwest stretching toward the beach at El Rosario. To the north and east is the granitic spine of Sierra San Pedro Mártir. Picacho del Diablo is the highest peak in Baja California. To its right is Pico Matomi, a Miocene volcanic neck of andesite porphyry. To the east is the Sierra San Miguel. To the south is Mesa la Sepultura.

VIEW TO NORTH OF ALISITOS ROCKS AND PICACHO DEL DIABLO

86   Along this ridge are beautiful views of Mesa la Sepultura. Mesa la Sepultura and Mesa San Carlos are composed of the Cretaceous marine Rosario Formation in the lower parts and are overlain by the Paleocene-Eocene marine Sepultura Formation (Santillan & Barrera, 1930). The Sepultura Formation consists of conglomerates which grade into mudstones with conglomerate channels to the west. This Cenozoic section extends inland 30 kilometers. It becomes thinner and more coarsely clastic and nonmarine with sandstone, conglomerate, and scattered thin shale partings that pinch against an irregular buried topography.

Due to runoff, exposure, and winds, life on this ridgeline is extremely harsh. The vegetational "cover" is sparsely represented by Cirio, Cardon, and Agave, all true xerophytes.

Xerophytes are plants. In Baja, xerophytes include succulents, such as various cacti or members of other families that have fleshy stems and leaves that enable them to store water for a long time. They frequently have shallow root systems and are able to utilize the soil moisture from light rainfall, heavy dew, or fog. Such plants take advantage of the little precipitation that falls in the desert areas of Baja by storing precipitation, dew, and/or fog in their pith and cortex parenchyma tissues for months or even years. Many succulents such as cacti have leafless and ribbed stems, two more adaptations for survival in desert environments. Leaflessness reduces the surface area through which water is lost by transpiration. The ribbed stem allows the stems to swell like an accordion and store water when it becomes available.

86.4 Exposures of reddish soils of the Alisitos Formation can be seen along the sides of the highway. There is a rather interesting little gorge in the Alisitos Formation to the right.

The vegetational cover consists of Teddy Bear Cholla, Barrel Cactus, Cirio, and variously colored crustose lichens that look like paint that has been "splattered" randomly on the rocks. Each color is a different lichen.

92.2 The side road is the turnoff that leads to the turquois mine, La Turquesa. The highway descends the new Aguajito grade through adamellite and the metavolcanic rocks of the Alisitos Formation.

LA TURQUESA: Turquoise, a light blue to blue-green phosphate mineral, is formed as a secondary deposit that fills cracks and shears in the metavolcanic rocks of the Alisitos Formation. Turquoise, which is the birthstone for December, usually occurs in cracks and as reniform masses with a botryoidal (the form of a bunch of grapes) surface in the zone of alteration of aluminum-rich igneous rocks. The La Turquesa mine became a series of "gopher (go for) holes" as the miners followed the illusive mineral. They mined what was absolutely necessary.

94 To the south is the bedded Alisitos Formation that dips slightly to the south. Excellent exposures of the Alisitos Formation, which is composed of rhyolite, ignimbrite, basalt, and andesite outcrop in this area.

SAND VERBINA FLOWERS IN THE SPRING

ABRONIA, the sand-verbenas or wild lantanas are perennial herbaceous plants. Despite the common names, they are not related to verbena or lantanas. They are native to western North America including Baja California growing on dry sandy soils.

LARGE CARDON AND CAROL IN 1967

94.4   The zigzag pattern of the old road is seen to the right.

95.6   The road crosses a flat valley on a Tertiary conglomerate that appears to have been an ancient river channel. This conglomerate may be correlative with Cretaceous age conglomerates formed when the area was first elevated and eroded due to the pushing of the Pacific plate under Baja.

Beavertail Cactus, Agave, Barrel Cactus, Pitaya Agria, Broom Baccharis, Toyon, sparse Cardons, and tall Cirios that are covered by *Ramalina* lichen, comprise the vegetational cover of this portion of the Vizcaino Desert.

101   The highway crosses a prominently jointed adamellite (looks like a dike).

105   This kilometer marks the beginning of a large Cardonal and Cirio forest. The vegetation of the area is typical of the Vizcaino Desert flora and is dominated by Desert Thorn, Mimosa, Mesquite, Desert Mallow (along the pavements edge), Cirio, Datilillo, Garambullo, Beavertail, Jumping Cholla, Barrel Cactus, Tamarisk (in washes), Yucca, Cardon, and Palo Estribo.

The highway passes Rancho El Arenoso, one of the numerous, old-time cattle ranchos of the peninsula. The old road led right to the front door of the ranch house where earlier Baja travelers often stopped for cold refrescos and a respite from the rough bone-jarring road.

Along the highway there are large barrel cactus that are known as the "compass plants" of Baja. Because of their tendency to grow toward the light of the sun to the southwest (to reduce the harmful effects of the sun during the hottest times of the year), they are called the "compass plant." The tops of the cactus are surrounded by reddish-thorns. In the late spring they have a ring of beautiful waxy yellow flowers. In an emergency the Barrel Cactus will yield approximately a pint of alkaline juice from its pithy cortical parenchyma (water storage tissues of the stem).

BARREL CACTUS

111   Cirio Forest.

DIP SLOPES WITH EROSIONAL CURVES

113   The highway roughly follows a strike valley parallel to the bedding with beds that dip toward the east at approximately 45 degrees in the metavolcanic rocks of the Alisitos Formation. In the hills ahead and to the right, two interesting erosional curves in the dipping limestone beds are visible. The beds are not folded.

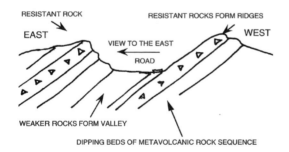

Prominent ridges that contain limestone are exposed on both sides of the highway. Look to the left at the long strike ridges (45 degree dip) across the valley in the near distance. This is reminiscent of the plateau area of Arizona and southern Utah, except the age of the rocks and the vegetational cover differ.

These **metavolcanic rocks** were formed in a Volcanic Island Arc as the Pacific Plate was subducted under the North American Plate. The large areas of metavolcanic rocks without fossil or radiometric dates are exposed throughout much of the Pacific slope of northern Baja. South of the Agua Blanca Fault, the volcanic sequence includes a variety of sedimentary strata. Limestone, calcareous siltstone, and mudstone are interbedded with volcanic sandstone, conglomerate, tuff, and breccia, and represent a wide variety of depositional environments. Deposits range from deep to shallow marine and nonmarine, coarse sedimentary breccia to limestone, and basalt to rhyolite. Andesite and andesite breccia are the predominant volcanic rocks. The entire sequence is thought to be tens of thousands of feet thick. It varies widely in stratigraphic components from area to area. No two measured stratigraphic sections are alike.

118   Yuccas begin to be more abundant.

YUCCAS are up to 12 feet high shrubs of the lily family with sharply pointed leaves and a large flowering stalk that supports thickly clustered, terminal panicles of whitish flowers. The flowers are pollinated by tiny night flying Pronuba moths. Neither the Yucca nor the moth can propagate its species without the aid of the other. The moth larvae feed exclusively on the Yucca seeds and the Yucca flowers are only pollinated by the moth. Eliminate one of the symbiotic mutualists and the other will die.

Both man and cattle make use of parts of the Yucca. The stalks are rich in sugar and the saponins stored in the roots are often used as a substitute for soap. (In fact, there was a shampoo in the early 70s called Yucca-Dew Shampoo. I was distinctive at the time for containing the oils from the Yucca plant. You can still find all-natural shampoos today that contain Yucca oils.) Saponins are plant glucosides that form a soapy colloidal solution when mixed and agitated with water. They are used in detergents, synthetic sex hormones, foaming agents, and emulsifiers. The flowers are edible. The cucumber-like fruits can be eaten raw or roasted. The long sclerenchyma leaf fibers were used in making baskets, sandals, and mats.

121.2 Turnoff to Misión San Fernando de Velicata (5.8 kms.) down the arroyo west of El Progresso. The Misión was founded in 1769 by Father Junipero Serra as he traveled north to Alta California. It is the only mission in Baja California established by the Franciscans.

123   The highway crosses a wash-woodland of Mesquite.

123.5 The highway ascends to the top of a small grade passing through the metavolcanic rocks of the Alisitos Formation. The old road went to the right, around the hill, and up the wash. The paved highway goes to the left, over the rise, and across the hill thus avoiding the floodwaters that run seasonally through the wash.

125   To the north is Valle El Renoso. To the south is the plain of Llanos de San Agustin.

PLEISTOCENE LAKE BED SEDIMENTS

A Pleistocene lakebed: The light-colored brownish-tan (buff) sandstone, siltstone, and limestone that occupy the bottom of Valle Santa Cecilia were part of one of a series of Pleistocene lakes in this part of the peninsula. Freshwater vertebrate fossils that include turtles have been recovered from the lake sediments. The sediments of this "fossil" Pleistocene lake are fine-grained, and this part of the highway is very dusty when dry.

127   The highway traverses to the east on the flat upper surface of the sediments on what was the lake bottom. On the north side of the highway, the shoreline of the lake is visible at the base of the low hills where it lapped against the metavolcanic rocks. In the distance, on the south side of the highway, the shoreline lapped against low rolling hills that are the fluvial part of the Paleocene-Eocene Sepultura Formation. The vegetation cover of the lake sediments is primarily Creosote Bush.

The vegetational cover is sparse and is represented by a few small Cirio, Cholla, Ocotillo, *Agave*, Creosote Bush, Yucca, and Mesquite. Against the base of the metavolcanic hills, the flora is dominated by bright green stands of Mimosa that delineate the course of the wash visible from Kms 127 to 140.

CREOSOTE BUSH is a common resinous desert shrub of the southwestern U.S. and Baja. Unlike many other desert plants, it is green year-round. The waxy coating on its small leaves reduces water loss by transpiration and allows the shrub to withstand long periods of extreme drought. The strong-smelling resinous leaves resemble the odor of the coal tar distillate creosote. Few plants grow under the Creosote Bush. The roots produce a toxin that inhibits the growth of other plants. After heavy or frequent rains, when the toxin is leached from the soil, the growth of desert annuals is permitted. As the soil dries, the inhibitor accumulates and poisons the outsiders. This phenomenon is known as allelopathy and it helps to eliminate the competition for water in a dry desert environment. Creosote Bush was considered a cure-all by Baja's early Indians, and early Spaniards used a decoction of this shrub to treat sick cattle and saddle-galled horses. In more recent times, a leaf extract is used to delay or prevent butter, oils, and fats from turning rancid. Creosote Bush leaf extract is not widely-used though as the compound is now synthetically manufactured.

Exhumed Erosion Surface This area consists of flat plains fringed by lava capped hills. The Llanos de San Agustin were eroded to a relatively flat surface by streams during the Early Tertiary. (The remnants of the deposits

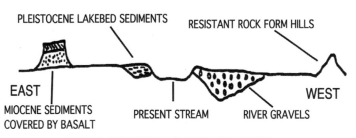

PLEISTOCENE LAKEBED SEDIMENTS   RESISTANT ROCK FORM HILLS

EAST   WEST

MIOCENE SEDIMENTS   PRESENT STREAM   RIVER GRAVELS
COVERED BY BASALT

STYLIZED SECTION OF LLANOS DE SAN AGUSTIN

from one of these streams can be seen as conglomerates on the road to Santa Catarina.) This base-leveled surface was then covered by Miocene fluviatile sedimentary rocks, basalts, and rhyolites that stretched across the peninsula. The area was uplifted sometime during the last 10 m.y. As a result of this uplift, erosion stripped the covering rocks and "exhumed" the Early Tertiary surface leaving lava-capped mesas scattered over the area as erosional remnants. Later, impounded drainage filled the low places with softer lakebed sediments that are being dissected by the present streams.

131   The buildings along the highway are the new Rancho Penjamo.

Several more of the old ranchos dot the old road to the left. The paved highway is approximately two kilometers south of the old road at this point. The highway is still following the flat surface of the Pleistocene fluvial lakebeds discussed at Km 125.

132.3 The side road to the right leads southwest to Santa Catarina and Puerto Catarina on the coast. The Santa Catarina road follows the coast for many kilometers to the south and finally rejoins the highway near El Tomatal (4:69.2). It is a fairly rough, but quite picturesque side road. It is only recommended for 4-wheel drive and group travel. Carol and I spent three weeks and four days mapping at the mouth of Arroyo San José. Only two parties passed us at that time and one of them desperately needed gas.

141   The flat reddish mesa ahead is Mesa Redonda; it is capped by volcanic rocks of Miocene age. Most of the flat hills in this area are mesas that are capped by these same volcanic rocks.

LAVA MESA ON EROSION SURFACE

142.3 The first Elephant Trees visible from the highway are growing on the slope to the right. These unique trees are discussed at 4:9.

149.9 Turnoff to the abandoned El Marmol Onyx Quarry (15 km) at El Marmol (marble). The road to the mine is usually graded and should be passable with reasonable clearance passenger vehicles.

EL MARMOL

EL MARMOL ONYX SCHOOL HOUSE

Hot mincralized solutions rise to the surface along a fault line, flow out on the surface, and evaporate as they cool. Travertine is deposited in layers. There is a large amount of travertine at the abandoned mine. The old onyx schoolhouse is still standing and an interesting graveyard is located nearby.

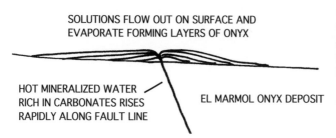

SOLUTIONS FLOW OUT ON SURFACE AND
EVAPORATE FORMING LAYERS OF ONYX

HOT MINERALIZED WATER
RICH IN CARBONATES RISES
RAPIDLY ALONG FAULT LINE

EL MARMOL ONYX DEPOSIT

EL MARMOL is the location of hot springs and a Travertine (onyx) deposit mined between 1900 and 1958 by Southwest Onyx and Marble Co.

150   The highway continues over Llanos de San Agustin. To the north (left rear 40 kilometers in the distance) are the two high peaks of the Sierra San Pedro Mártir: the granitic pluton of Picacho del Diablo,is the highest peak in Baja California at 10,124 feet, and Pico Matomi, a Miocene volcanic core of andesite porphyry, to its right. To the far right (west) are irregular hills of the metavolcanic Alisitos Formation.

For years, there has been a Red-tail Hawk nest in the large Cardon just to the right of the highway. Careful observation of other large plants along the may reveal other nesting sites of predatory birds.

PREDATORY BIRDS OF BAJA: Over 270 species of birds of prey hunt in the daylight in the world. Over 140 species of owls are the world's nocturnal predators. In Baja, there are 16 daylight and 4 nocturnal species of predators.

Predatory birds exhibit numerous anatomical and behavioral adaptations that enable them to lead a predatory life-style. For example, birds of prey generally have long curved talons for seizing their victims and strongly hooked beaks for tearing them apart. They are masters of soaring and swooping (diving). Baja's Peregrine Falcon may reach 175 mph as it dives for its prey. In certain hawk species, the eyes are larger and more acute than those of humans; they are binocular instead of being set on the side of the head as in most non-predatory birds. The most commonly seen predators along the highway are Red-Tail Hawk *(14:167)*, Red-shouldered Hawks *(here)*, Northern Harriers (Marsh Hawks) *(12:145)*, Ospreys *(4:128)*, Peregrine Falcon *(8:25)*, and Turkey Vulture (Buzzards) *(8:57.2)*. Each will be discussed at kilometer marks where they have been commonly seen.

The **Red-shouldered Hawk** is distinguished by its reddish shoulders, and narrow white bands on the wings and tail. Viewed from below, their under parts and wing linings are uniformly reddish. Their flight is accipiter-like with several quick wingbeats and a glide. Red-shouldered hawks inhabit mixed woodlands and are often seen near streams. They hunt from a perch for snakes, frogs, mice, and young birds.

153   The prominent flat-topped mesa to the right is Mesa las Palmillas; it is covered with beautiful stands of Elephant Trees, Ocotillo, and the weird Cirios.

155   The Cardon at the end of the mesa to the right has watched over travelers for decades.

158.5 Some of the first bouldery granitic rock outcrops, of the Las Virgenes region are seen under the volcanic-capped mesas, visible to the right of the highway. This region, known as Las Virgenes, is full of large, spectacularly picturesque granitic rock formations and many varieties of cactus and other desert vegetation. This region is also known as the Cataviña Boulder Field.

159   Notice the Barrel Cactus slanted toward the southwest in an effort to reduce the harmful effect of prolonged exposure to the southwestern suns rays during the hottest months (*See* 3:106).

162   The vista opens to reveal numerous lava-capped mesas with cinder cones on top of the picturesque bouldery outcrops of the Las Virgenes area.

SPHEROIDAL   WEATHERING can easily be explained. This area receives little rainfall; weathering proceeds slowly on the surface of the rocks. Since there are three surfaces at corners and two at edges, the rocks tend to weather into spheres. The weak running water carries the finer particles away which leaves the rounded boulders.

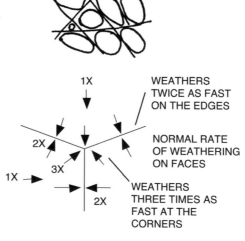

1X   WEATHERS TWICE AS FAST ON THE EDGES

NORMAL RATE OF WEATHERING ON FACES

2X   3X   1X

WEATHERS THREE TIMES AS FAST AT THE CORNERS

2X

167.8 This side road goes to La Bocana and Rancho San José (95 Km). Get a guide and follow Arroyo La Bocana to the southwest for views of pre-historic painted Indian rock art (pictographs). A display at the Parador gives some information about the "Zona Arqueologica" in this region. Rock art can also be seen in Arroyo El Palmarito (Km 170). There is a view of the bouldery granitic tonalite terrain with basalt-capped hills of the Las Virgenes region. The peaks to the right are the Alisitos Formation. The metavolcanic rocks have yielded fossils from the Boca San José area.

ELEPHANT SEAL ROCK AND CREOSOTE

The vegetation is typical of the Vizcaino Desert flora consisting primarily of Cirio, Cardon, Elephant Trees, Garambullo, Cholla, Creosote Bush, Barrel Cactus, annual composites, Teddy Bear Cholla, an occasional Creosote Bush, Ephedra, *Agave*, Jojoba (goat nut), Pitaya Dulce, Pitaya Agria, and scattered Acacia, Datilillo, and Burro Weed. The orange "hair-like" plant that grows parasitically on the Elephant Trees is Dodder (a.k.a., Witches Hair). The opportunistic Brittlebush is on the highway edge. Smoke Trees and Mesquite are restricted to the washes.

There is evidence of mammals feeding on cactus. This feeding on cactus obtains metabolic water as well as food. Most Cardons bear scars from wood rats below 15 feet. Barrel Cacti are commonly hollowed out by small rodents and jackrabbits. Over half of the Chollas are fed upon by jackrabbits.

ARROYOS are washes cut by intermittent streams and flash floods. Baja's arroyos, which are normally dry, are infrequently filled with water after it has rained. Flash floods fill the arroyos from wall to wall with violently moving, sand-laden waters. The washes may run for several days depending on the amount and duration of the preceding rains. After these flash floods, the arroyos will be as dry as if they had never conducted a single drop of water.

The vegetation is denser in the Cataviña area. The boulders shed rain to provide water for plants near the boulder, thus increasing the net rainfall.

171   This area is considered by many travelers to be of the most scenic in Baja due to its bouldery beauty and the starkness of the Vizcaino Desert

Region and its vegetation. Picturesque rock formations and flora are found along this stretch of the highway. Good camping is available especially along part of the old road that winds close to the highway.

CATAVIÑA BOULDER FIELD

Birds from the California Phytogeographic Region are seen here due to the proximity of that region as well as the higher moisture produced by the rocks. A winter study produced the following bird list for the Cataviña area. More birds were seen in rocky sites as there is more protection from wind, more diversity of perches, and more abundance and variety of plants.

BIRDS OF THE CATAVIÑA REGION

*Anna's Hummingbird	Ash-throated Flycatcher
Black Phoebe	*Black-chinned Hummingbird
Black-throated Sparrow	Black-tailed Gnatcatcher
Blue-gray Gnatcatcher	Brown Towhee
Cactus Wren	California Quail
Cassin's Kingbird	Common Raven
*Costa's Hummingbird	Gambel's Quail
*Gila Woodpecker	Gilded Flicker
Phainopepla	Ladder-backed Woodpecker
Mockingbird	*Red-shafted Flicker
*Roadrunner	w - Robin
Rock Wren	Rod-tailed Hawk
Say's Phoebe	w Scott's Oriole
Sparrow Hawk	Turkey Vulture
Verdin	w - White-crowned Sparrow
White-throated Swift	
* courtship or nesting activity	w - possible wintering species

PALM AT CATAVIÑA

176   The highway descends into Arroyo Cataviña (Arroyo El Palmarito). To the left, flat-lying volcanic rocks cap the erosion surface on the granitic rock.

Take some time in this area to explore and take pictures. Dawn, dusk, and afternoons are particularly picturesque photographic times. Adopt the motto of naturalists everywhere: "take only pictures and don't even leave foot prints" so that others may continue to enjoy.

176   Along the left side of the highway a short distance up Arroyo Cataviña is a large lush stand of native palms; they include specimens of fan palms (*Washingtonian falifera*, and *W. robusta*), and blue fan palm (*Erythea brandegeei*). Palms in the desert often indicate perennial water.

NATIVE PALMS OF BAJA: Native palms were important to Baja's Indians. They ate both the thin fleshy portion we normally eat as well as the large seed. Palm fronds made numerous useful products such as sandal-like footwear, baskets, ceremonial effigies (dolls) of the dead, and house and roofing materials. There is evidence that the Indians also burnt the trees periodically to kill insects and mites and to improve the date yield for the next harvest. The palm seeds are also eaten by birds and other animals. Orioles use the fibers from the leaf for nesting material. Other vegetational dominants of the area are large Dodder-covered Elephant Trees, Barrel Cactus, Cardons, and Cirios.

Archaeological evidence indicates large numbers of prehistoric native Indians once occupied the area and made use of the perennial water located in this region. In boulder areas west and north of Parador Cataviña, stone chips can be seen littering the surface of the ground. Inquire at the Cataviña Hotel or at Santa Ines for a description of how to get to the pre-historic rock paintings (pictographs) located. southwest in Arroyo La Bocana. These pictographs are known locally as "Cueva de las Pinturas Rupestres Gigantes."

HISTORY OF MAN IN BAJA: The early history of man in Baja is not well known. The author of this guide has found, but not disturbed, many archaeological sites over the past 50 years. Scientific work is proceeding on the peninsula and on the Gulf islands. As one author wrote, even the islands remain "...largely an archaeological terra incognita." Extensive excavations and mapping of human habitation of the peninsula remains to be done to illuminate the history of human occupancy of Baja. Most of the investigations of human occupancy that have been completed so far are either site reports or studies of artifact collections.

The exact origins of the prehistoric Indians of the North American continent is unknown. One of the most popular theories of the origin of the American Indian states that millennia ago, man came from Asia across the Bering Strait of Alaska either by way of the Diomedes Islands (whose formation served as a ladder) or the Aleutian Islands. It is thought that the various races of Indians arrived at different times and in several different migratory surges. All that is known for certain is that man has long inhabited the peninsula as evidenced by the painted caves, petroglyphs, pottery shards, spear points, scrapers, manos and matates, debitage, carved bone, stone circles, rock cairns, lithic knives, shell middens, fragments of Yucca and Palm sandals, baskets, ceremonial effigies, and living structures.

176.5 The highway passes through a low saddle and descends into Arroyo La Bocana. Pools and water down the arroyo can be seen most of the year. Downstream to the west on the "nose" of a large granitic mass, large, smooth, water-polished boulders are visible 25 feet above the wash bottom. At times, there must have been at least 25 to 30 feet of water in this arroyo to have polished the tops of these enormous boulders.

It is possible to travel up nearby Arroyo La Bocana, 23 kilometers, to Misión Santa María. It was founded in 1767 by the Jesuit Priest Father Victoriano Arnes. If you do not have a 4-wheel drive vehicle, it is worth spending some time in Santa Ines and have someone take you up the arroyo to Mission Santa María to explore this area.

This stretch of the highway passes directly over the site of the popular gas stop of old Rancho Cataviña. Gas was siphoned into 5-gallon cans and then strained through a chamois or a felt hat into the gas tank.

179 Hotel Mission de Cataviña.

180.2 Shortly after leaving El Parador, the highway crosses the Arroyo Santa Ines with its perennially-running stream and another grove of native fan and blue palms. There is a picturesque view upstream of the rocks and another native palm oasis.

SPHEROIDAL GRANITICS, BLUE PALMS AND LAVA MESA

180.8 The paved turnoff to the left leads 2 Kms. to Rancho Santa Ines and a paved airstrip.

182   Elephant trees are the dominant plants.

185   The grade climbs through Miocene fluvial sedimentary rocks that stretch across the highway in a wide band. Fluvial sedimentary rocks are overlain by basalts similar to those on Llanos de San Agustin.

187   A conical hill on the lava-capped mesa can be seen on the skyline at to the south. It represents the eroded remnants of a small basalt cinder cone.

188   This section of the highway passes through granitic rock overlain by Miocene volcanic rocks. Other conical-shaped hills, that also represent cinder cones, are in view on the horizon ahead.

189   White dikes cut across the mixed granitic and metamorphic rocks between here and Km 196.

191   Native fan and blue palms are growing to the right in the nearby Arroyo Jaraguay. The vegetation of this area is denser than that of other parts of the

Vizcaino Desert because of more moist soil. The vegetational dominants are typical of the Vizcaino Desert flora as represented by bright green Mimosa trees, Tall Cardon and Cirio, and an undergrowth of Garambullo, Pitaya Agria, Pitaya Dulce, Brittlebush, Creosote Bush, Ocotillo, Atriplex, Jumping Cholla, scattered Palo Estribo, and species of annual composites.

191.5 Rancho San Martín. The "golden spike" was planted here and commemorates the completion of the Transpeninsular Highway in 1973.

196.5 Turnoff leads southwest to Rancho Jaraguay and the old road. Just before the ranch on the left are the ruins of a number of adobe buildings.

Outcrops of gneisses and tonalite with andesite dikes are exposed along the highway. This region is one of impressive mixtures of metamorphic and granitic rocks, dikes, bouldery outcrops, and beautiful basalt-capped mesas such as Mesa Jaraguay, Mesa El Gato, and Mesa Prieta.

198 The highway climbs a steep grade and crosses over the old road.

199.5 As the highway crests the top of Jaraguay grade (2700'), a good view opens to the north (rear). This viewpoint provides an overview of the volcanic tableland the highway has been passing through, the unconformity between the granitics and the basalts, and the relatively flat erosion surface that underlies the volcanic strata. A few Cirios are growing on the drier south side of the grade. Many larger Cirios can be seen on the moister north side.

NORTH AND SOUTH FACING SLOPES: Local variations of humidity and temperature exercise considerable control upon natural slope cover. In nontropical regions, slope face is of great importance. Because Baja is north of the Equator, its southward-facing slopes receive more direct sunlight, thus, they are hotter and drier. As a result of the decreased humidity and increased temperatures, slope cover is sparse, low, and composed of grayish-colored species that are adapted to living in xeric (dry) environments.

North facing slopes receive less direct sunlight and are more humid and cooler than south facing slopes. They support denser, taller, moisture-needing vegetation including trees characteristic of mesic (moist) environments. Wind is slowed, and soil temperatures and evaporation are decreased. The soils are less weathered and are humus-rich.

200.5 View of Laguna Seca, a normally dry lake, to the left. The old road used to bear to the left and crossed part of the lake bottom. At the present, the highway circles around to the right side of the lake on the high ground near the base of the hills.

203.5 A view opens to the south. Several volcanic peaks of the San José

93

volcanic fields can be seen in the distance to the right. The rolling landscape in this region is covered with volcanic debris.

204.9 The highway crests a rise for the first good view of Cerro El Pedregoso.

210 The highway passes close to Cerro El Pedregoso. The base of this granitic hill has been covered by volcanic and other debris so that only the top stands out like an iceberg above the volcanic plain (Pliocene-Holocene basalts and basaltic andesites). The granitic rocks have been spheroidally weathered. This feature on the otherwise fairly barren, gently rolling landscape acted as a beacon for Baja travelers; they could see for many kilometers as they slowly approached the hill over the old rugged dirt road.

EL PEDREGOSO.

220 This road cut exposes andesitic dikes in tonalite. In the late spring, the tall "weeds" with yellow-centered white flowers that grow along the roadside are called California Prickly Poppies.

221 The red and black pyroclastic cinders and basalt exposed in this road cut indicate that an ancient eruptive center is located nearby.

222 A prominent white quartz dike is exposed 300 meters to the left.

226 The hill ahead is cut by prominent, gray-white, granitic dikes.

227 The highway begins a gentle descent as the dry playa lake of Laguna Chapala comes into view. The mountains to the southeast are part of the Sierra la Asamblea. The valley in front of them is Valle Calamajué. The vegetation is mostly Creosote with scattered Cordon.

The old road went to the left across the lake to a small group of buildings and trees. Now, the highway skirts the west side of the lake between the upper and lower lake at the New Rancho Chapala.

At Laguna Chapala, there is a vegetation change. The Cirio almost disappears; Cardons become very small against the foothills of Sierra de Calamajué. The vegetation adjacent to the highway is dominated by low gray-mounded Bursage and Creosote. Datilillo, Cholla Cactus, Pitaya Agria, Desert Mallow, and Teddy Bear Cholla are scattered among the halophytes.

233.5 The road to the left leads to the northeast to Bahía de San Luis Gonzaga, Puertecitos, and San Felipe. (*See* Log 11).

234.8 Ejido Sierra de Juárez.

TILTING OF THE PENINSULA? The old shorelines of Laguna Chapala are tilted relative to the Recent shorelines of the lake. This indicates Pleistocene to Recent tilting of this area.

236 The semi-stabilized dunes on the south end of Laguna Chapala are blown from the lake during normal dry spells. A climb to the top of the dunes will provide an excellent view of Laguna Chapala.

239.7 The hills ahead of this kilometer mark are composed of tonalite that is cut by dark andesitic and basaltic dikes that fed the extensive basaltic volcanic in the hills to the right. There are excellent examples of black basaltic dikes cutting granitics.

INDIAN TREE TOBACCO is the common name of the tall (3-10 foot) herb that grows in the disturbed soils along the roadsides throughout the peninsula. This plant is easily recognized by its yellow tubular flowers and disagreeably strong scented poisonous narcotic leaves. It is a weed that was introduced from Argentina and Chili. It is found the length of Baja and in the deserts of the Southwest. This plant is reported to be poisonous to stock.

243.2 The highway crests Cuesta El Portezuelo. This is the main peninsular divide where the highway passes over a ridge from the Pacific Coast drainage to the Gulf drainage. Climb the hill for a spectacular view and photo op.

CUESTA EL PORTEZUELO. The old and the new highway pass through here.

The highly mineralized granitic and metamorphic rocks of Sierra la Asamblea form the high mountains to the left front. The high light-gray part of the Sierra is a granitic pluton that is surrounded by metamorphic rocks. Several light-colored dikes cut the darker metamorphic rocks.

SIERRA LA ASAMBLEA

246 The vegetation of this area is similar to the granitic boulder gardens that surround Cataviña and is dominated by Cardon, Cirio, and Elephant

Trees, Old Man Cactus, Teddy Bear Cholla, Yucca, Agave, Jumping Cholla, Pitaya Agria, Palo Adan, and Ocotillo.

252.5 There is a view of Cerrito Blanco directly ahead. It is a low light-gray, highly jointed granodiorite hill that sits on an alluvial plain that is unconformably overlain by white cross-bedded Paleocene sandstones. The jointing makes the hill look like it is composed of bedded sedimentary rocks.

255 The light-colored flat Pleistocene lakebed sediments are exposed along of the highway for the next several kilometers.

View to the north of a volcanic tableland and the cinder cone peak of Cerro el Volcancito.

AERIAL VIEW OF FAULT forming north edge of the stable San Borja block.

261 The highway crests a very low divide and moves back into the Pacific slope drainage.

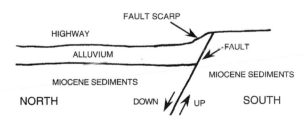

265.1 This small grade represents a prominent fault scarp that cuts diagonally across the highway. The fault uplifts Miocene fluviatile sedimentary rocks on the south against the Quaternary alluvium and lakebed sediments on the north. This is the north

97

edge of the stable San Borja block (Gastil *et al.*, 1975). It consists of a pre-Miocene, west-sloping bedrock surface that is discontinuously overlain by Cenozoic sedimentary and volcanic strata. South of here, the uninterrupted mesas extend from the main Gulf escarpment to the Pacific coastal plain.

262   The old road turns off to Bahía de San Luis Gonzaga, Puertecitos, and San Felipe. This road was used as a route to reach the newer road from Laguna Chapala at Km 233.5. This road passes by Misión Calamajué and Cerro el Volcancito through Valle Calamajué between Sierra la Joséfina and Sierra Calamajué, and connects with the graded road to Bahía de San Luis Gonzaga.

Several more raptor nests can be seen in Cirios along the highway. Woodpeckers commonly excavate holes (nests) in the Cardons in this area. Wood rat nests can also be seen in the centers of Yucca clumps.

268   The highway begins a gradual descent into Arroyo el Crucero. This drainage is developed on and eroded into Paleocene marine and nonmarine sedimentary rocks.

Just north of Punta Prieta, the dominant species alternate between Cirio, Datilillo, and Cardon. The low sparse gray "understory" is Bursage, Agave, Garambullo, Jumping Cholla, and Acacia. This Datilillo-Cardon/Datilillo-Cirio community continues for many kilometers.

HOW DO THE LOW SPARSE GRAY FLORA OF BAJA'S EXTREMELY ARID DESERTS SURVIVE? Most of Baja's desert landscape is dominated by low-growing vegetation (shrubs, cacti, and grasses) although some plants do grow to the size of small trees. Some are leafless (obligate drought-deciduous species) most of the year. In the absence of leaves, the green trunks (cladodes) act as photosynthetic leaves and produce carbohydrates to meet the plants required energy resource. Others have silvery or velvety foliage that helps reflect heat and retain moisture under the hot desert sun. Many are armed with spikes or thorns, to discourage animals from grazing on their slow-growing branches. The spiked leaves of many plants provide points from which dew can collect, fall to the ground, and water the plant. Other adaptive features are the brief life span of some plants that germinate, grow flowers, and produce seeds in days to a few weeks and the storage of water in tissues and the accordion structure of cacti that swell to hold more water. Some have very large trunks like the Elephant Trees or develop thick and waxy outer skins. Wide spacing, shallow roots, and vegetative reproduction are also water saving adaptations.

282.1 Junction of Highway 1 and the highway to Bahía de Los Ángeles (Log 16). The kilometer markings on the highway change to 0.

Log 4 - Bahía de Los Ángeles Jct. to Guerrero Negro [129 kms= 80 miles]

L.A. Bay Turnoff to Rosarito - *The highway continues south in the Miocene fluvial sedimentary rocks with basalt mesas to the left and high rugged metamorphic and granitic hills to the right. It then drops into the main alluvial channel of Arroyo Leon, past Punta Prieta, with Paleocene fluvial sedimentary rocks on both sides of the arroyo and metamorphic and granitic hills in the distance. At La Bachada, the highway climbs up a grade through rolling hills onto the rolling Paleocene mesas with isolated steep hills of metavolcanic rock and high volcanic mesas in the distance. After passing a rugged metavolcanic hill to the left, two conical shaped Occidental Buttes composed of Paleocene fluvial sedimentary rocks are seen to the right. The highway then enters a hilly area of granitic rocks and drops into El Rosarito.*

Rosarito to Guerrero Negro - *The highway follows the south bank of the arroyo with metamorphic rocks on the left, then passes through a canyon in the steep gabbro hills before turning south to again climb onto and through rolling hills in the undulating dissected mesas of marine Paleocene sedimentary rocks. The highway drops into a steep narrow gorge in gabbro then alternately climbs onto the Paleocene marine mesas and drops into the valleys with the basalt capped Paleocene mesas to the east. Finally, the highway drops onto the alluvial fans of the Llano del Berrendo. During the long crossing of this plain, the low hills of metasedimentary rocks to the left become more distant. A basalt cone is closely passed to the right and the distant basalt cone of Punta Santa Domingo looms ever closer. The coastal dune field and the Pacific Ocean are in almost constant view to the right. After passing Jesus María, the highway begins to cross the limestone surface of the Pleistocene lagoon. The dune field encroaches on the highway as the Eagle Monument is approached at the state line.*

0   The highway to the left leads to Bahía de Los Ángeles (See log 16) while the highway ahead continues 129 kilometers south to Guerrero Negro.

4   The flat mesa to the left front is Miocene fluvial sedimentary rocks capped by basalt.

6.5 The Cardons that are seen along this stretch of the highway are reputed to be among the tallest in Baja, even taller than some specimens in the large cardonal located at the south end of Bahía Concepción on Baja's gulf coast.

9   The dominant trees along the highway are two unrelated genera of Elephant Trees, Pachycormus and Bursera.

ELEPHANT TREE IN BLOOM

The name **ELEPHANT TREES** is unfortunately applied to several unrelated desert trees in Baja. The two kinds of Elephant Trees in this region are not closely related at all. *Pachycormus discolor* is a member of the Cashew family, while *Bursera microphylla* belongs to the Torchwood family. These two Elephant Trees are easily differentiated because of the distinctive incense-like aroma that is given off by the crushed leaves of the *Bursera*.

Bursera microphylla ranges from the Anza-Borrego Desert in the U.S. through the entire desert regions of the peninsula to the Cape region. A second species of *Bursera, B. hindsiana*, has almost the same distribution. Oil from the fruit of *Bursera* has been used as a dye and for tanning hides. A fourth species of Elephant Tree, *Bursera odorata* may also be seen.

13.2 This turnoff leads to Punta Prieta (named for the dark basalts in this region). The highway follows Arroyo Leon which is developed along a fault that is responsible for the arroyo's north-south trend. The Paleocene Sepultura Formation is exposed on both sides of the arroyo for the next 10 kilometers (*See 4:29*).

14 The airstrip at Punta Prieta is to the left. The hills to the right are composed of metavolcanic rocks of the Cretaceous Alisitos Formation. The Paleocene sedimentary rocks lapped against the irregular topography of the Cretaceous hills. The high hill to the east is composed of tonalite.

OCOTILLO OR PALO ADAN? These plants along with the Cirio belong to the Ocotillo family, a group of shrubs with long, erect thorny whip-like branches. It may seem like Ocotillo (*Fouquieria splendens*) and Palo Adan (*Fouquieria diguetii*) are the same plant. The characteristics that help separate these two close relatives are trunk, branch, and flower morphology (external appearance); geographical distribution; and habitat preference. The chart will help to distinguish between these two look-alike relatives.

	Palo Adan	Ocotillo
Trunk	short and thick	none
Branch diameter	Thick; branches off a short trunk	Thin, slender, whip-like, spreads upward, fan-like from ground
Flowers	Panicles smaller	Panicles larger
Geographical distribution	Abundant from the central part of the peninsula south to the Cape	Abundant from California to just north of Guerrero Negro rare south to the Cape
Habitat	Clay and granitic preference and plains	Desert slopes soils of alluvial plains

PALO ADAN   **OCOTILLO**

16.5 The denser vegetation in this region continues to be representative of the flora of the Vizcaino Desert. The plants along the highway are predominantly Cardon with occasional Palo Verde, tall lichen-draped Cirios, Agave, infrequent shorter Barrel Cactus, Datilillo, Pitaya Agria, Pitaya Dulce, two species of Elephant Trees in the washes, Jumping Cholla, Teddy Bear Cholla, Creosote Bush, Broom Baccharis, Cheesebush, and the last scattered occurrences of Ocotillo. From this point to the south an Ocotillo look-alike, Palo Adan, will begin to replace Ocotillo.

24.5 The highway turns and crosses Arroyo Leon where a wash-woodland is dominated by Mesquite and scattered Datilillo.

25.5 This is Rancho La Bachada whose well was once one of the better sources of water in this part of the peninsula. In 1965, the San Diego State University geologic mapping crews were based at Punta Prieta. Once a week, someone was sent to La Bachada to get 100 gallons of drinking water from the well. One week, a crewmember was sent to fill eighteen 5-gallon jugs. As he was filling the eighteenth jug, he lifted a bucket of well water that contained a dead decaying hairless rat. Not wanting to go elsewhere to refill the jugs, he pitched the rat, drew another bucket of water to fill the last jug, and returned to camp without saying a word (he used the other two 5-gallon jugs that week). No one found out until years later.

26   The highway begins a steep ascent through Paleocene conglomerates at the northernmost edge of a Paleocene embayment. Elephant trees are the dominant plant.

29   At the top of the grade, the vista opens to the south and presents a view of the Pacific coast and the flat Paleocene Sepultura Formation that forms the flat-topped mesas in the foreground and into the distance. The rugged high peaks to the far right and the high sharp peak directly ahead are composed of metasedimentary rocks of the Alisitos Formation.

This area was once a Paleocene marine embayment. Shallow water marine to nonmarine sedimentary rocks from this embayment have yielded fossils of hackberry seeds, marine invertebrates, a tooth of the proto-horse *Hyracatherium*, and non-marine vertebrate fossils (by Occidental College).

THE PALEOCENE ENVIRONMENT: Fife (1968) reconstructed the probable Paleocene environment and stated, "The coast north of Punta Santa Rosalía... was rocky as it is today. The location of fossiliferous deposits indicates that heavy-shelled gryphoid oysters were deposited with conglomerates near shore; while turritellas and echinoids were preserved with finer clastic sedimentary rock in the shallow embayments..." "South of Punta Santa Rosalía at least three major embayments existed..." "The southern embayment ... contains shallow to brackish water faunas. These were areas of mud to sand bottoms. The position of fossil mollusca, corals, ostracodes, and foraminifers suggests that alternating shallow marine and estuarine conditions existed. Seeds of the Family Chenopodieaceae and the remains of the proto-horse *Hyracatherium* are found in beds interfingering with typical estuarine strata. Ostracodes and charophytes indicate a lacustrine environment in the beds east of Rancho La Bachada."

PALEOCENE SEPULTURA FORMATION: "The most extensive outcrop of fossiliferous Paleocene marine rocks occurs in the southern part of the area, in the vicinity of and to the north of Rancho San Javier. In this region at least 100 feet of dusky yellow to reddish brown sandstone is interbedded with conglomeratic sandstones, concretionary lenses, siltstone, and mudstone. Several resistant horizons contain *Turritella pachecoensis* almost exclusively. One locality yielded *Cerithidea sp.*, *Ostrea sp.*, *Venericardia sp.*, *Glycymeris sp.*, and *Turritella pachecoensis*, plus several specimens of gryphoid oyster and unidentified horn corals, ostrocodes, and foraminiferas. This assemblage represents a sublittoral facies. Farther north, ten kilometers, up the arroyo from El Muertito, the assemblage suggests a lagoonal or littoral environment. with the remains of a small proto-horse, *Hyracatherium sp. nov.*, interbedded with seeds from the family *Chenopodiaceae, Cerithidea sp.*, *Calyptraea sp. nov.*, and *Ostrea sp....*" (Fife, 1968). "About ten kilometers due north of the above locality, at Occidental buttes, Morris (1966) reported the discovery of ungulates of the Orders Tillondontia, Perissodactyla and Pantodonta. Pantodonts of the family Barylambdidae were found stratigraphically above specimens assigned to Tillodontia and Perissodactyla..." (Fife, 1968).

THE ORIGIN OF THE HORSE AND ITS PLACE IN BAJA TODAY: It's hard to believe, but horses evolved in North America. Modern domestic horses (*Equus caballus*) belong to a small order of mammals known as the Perissodactyla or odd-toed ungulates which first appeared in the late Paleocene (58 million years ago) in North America. The earliest of the horse-like ancestors *Hyracatherium* appeared in the Eocene about 54 million years ago. *Hyracatherium* was a small dog-sized mammal that browsed on low shrubs of the forest floors. It had already lost two hind toes on its hind feet and one on its forefeet, but the feet were still covered with soft pads.

When grasses appeared in the early Miocene, hypsodont-grazing equids appeared. By the late Miocene, equids had reached their peak of diversity. The need to run from predators and to travel long distances in search of food and water led to many changes in equid body shape including increased body size. By the early Pleistocene, the one-toed equids (*Pliohippus*) had spawned the genus *Equus* that spread rapidly.

The center of equid evolution was North America. True equids did not migrate to the Old World from North America until the early Pleistocene about two million years ago. The Old World equids evolved from the three-toed North American immigrant equid *Hipparion*. By the end of the Pleistocene, modern equids had become widely distributed on all continents except Australia. For an unknown reason all North American ungulates, except the pronghorn antelope, became extinct about 10,000 years ago. In the Old World, the modern horse, *Equus caballus*, was first domesticated in Asia about 5,000 years ago. The domestic horse was reintroduced into North America (mainland Mexico) by the Spanish in the early 16th century.

32.5   This hogback ridge is composed of complexly folded and overturned metavolcanic rock of the Alisitos Formation. This unit is exposed to the left (east) for the next 4 Kilometers.

39   The small palm oasis to the right is Agua de Refugio. The waters of the small stream originate from springs located to the left of the highway. The highway crosses a small heavily vegetated wash with a perennial stream that originates in the metavolcanic hills to the left.

OCCIDENTAL BUTTE

44   Occidental Buttes are two small buttes in the Paleocene Sepultura Formation. This is the site of vertebrate fossil discoveries by Occidental College in the mid-1960s. Mother nature is taking a sunbath.

45   The vegetation of this region continues to be dominated by species characteristic of the Vizcaino Desert, and Cirio is the predominant tall plant. Plants associated with the Cirio are Barrel Cactus, Elephant Trees, Purple Bush, Agave, Pitaya Agria, with few Cardon, Palo Adan, and Jumping Cholla. *Ramalina reticulata* can be seen growing on the Cirio and Elephant Trees.

46   The highway climbs through conglomeratic beds and begins a descent through red beds of the Paleocene Sepultura Formation.

47   The high mesa to the left of the highway is the lagoonal section of the Paleocene Sepultura Formation overlain by basalts.

49   In addition to the lichen, Ball Moss is growing epiphytically on the Cirio and Elephant Trees.

EPIPHYTES AND PARASITES: From south of El Rosario along the Pacific coast to the Cape Region, trees, shrubs, and cacti such as Palo Adan, Cirio, Lomboy, Broom Baccharis, Cardon, and Pitaya Dulce are often abundantly covered with the moisture-loving epiphytic Ball Moss, a member of the pineapple family and *Ramalina*, a foliose lichen. The epiphytic Ball Moss lives in a commensalistic relationship and the Epiphytic Ramalina lives in a mutualistic symbiosis as a combination of an algae and a fungus. Both plants depend only on the host plant for support and are not harmful to the host plant. However, a close look at the same trees, shrubs, and cacti listed above will reveal harmful parasites that live in a destructive symbiosis with them. The two common parasitic organisms of this area are the evergreen mistletoe (*Phoradendron californicum*) and "witches hair" (*Cuscata veatchii*).

WITCHES HAIR is a plant that lacks chlorophyll, attaching yellow or orange hair-like stems to its host by a modified stem called a haustoria. Using the haustorial stem, it steals sugar produced photosynthetically by its host.

The second parasite, the **evergreen mistletoe**, hangs in trailing shaggy strands from many plants in Baja. Mistletoe also utilizes haustorial stems to drain sugar from its host. The moisture that sustains both the epiphytes and parasites of this desert comes from ocean fogs (Neblina) that roll in at night and lies during the early morning hours in misty layers among the hills until the sun "bakes" it off. Along the Pacific from the Cape north to San Francisquito, similar areas are known as coastal fog deserts. Mistletoe is less common on the drier Gulf side of the peninsula where fogs do not occur.

105

RAMALINA is an epiphytic foliose lichen formed by the combination of algal cells that live inside the tissues and cells of a fungus in a mutualistic symbiosis. The algal cells provide sugar to the host fungus that utilize it as a source of energy. In return, the fungus provides protection, and the water and carbon dioxide utilized by the algal symbiont during photosynthesis. Other species of lichen form living crusts on rocks in this area. Crustose lichen look like variously colored "splashes" of paint. A lichen is a symbiotic relationship between an algae and non-photosynthetic fungus. Due to a very specialized life system, they can exist on bare rock and obtain food from the air, sunlight, rainwater, and rocks. They are instrumental in the breakdown of rocks by their production of weak organic acids that result in the production of soils. The lichen also keep the rock wetter and promote more local chemical weathering. Lichens are common desert dwellers and are often the dominant vegetation in certain desert ecosystems.

52.5   The main part of the small village of El Rosarito is located to the left.

55   The wash that roughly parallels the highway is vegetated by green grasses, rushes (*Juncus sp.*), and scattered native palms.

After crossing the arroyo, the low hills to the right of the highway are gabbro. The hills ahead and a bit to the left are pre-batholithic slates.

A rough road from the center of town leads east 32 kms through very dense vegetation to Misión San Borja. The mission was completed in 1762 with money supplied by María, the Spanish Grand Duchess of Borja. It was abandoned in 1818, but has since been restored. It is a beautiful place to visit with an off-road or high clearance vehicle.

MISIÓN SAN BORJA

MISIÓN SAN BORJA NEOPHITE QUARTERS

56.8   The highway turns south out of the arroyo. Granodiorite is exposed to the right in the lower end of the Arroyo El Rosarito.

59   Beds of the Sepultura Formation are exposed on both sides of the highway. The low hills to the east are schist and gabbro.

66   The highway descends into a gabbro pluton exposed in the narrow neck of a canyon and then climbs through the reddish beds of the Paleocene Sepultura Formation.

The southernmost Cirios are growing here with Elephant Trees, Candelabra Cactus, Detillo, Pitaya Agria, Palo Adan, and Creosote Bush.

68.5 The highway descends into the broad arroyo. As it climbs out of the arroyo, the highway passes through outcrops of the Sepultura Formation. The prominent red hill directly to the left is a gabbro-serpentine pod.

69.2 The road to the right leads to El Tomatal Beach and Miller's Landing. Onyx from the abandoned mines at Marmalito was once shipped from the white-water beaches of El Tomatal.

70   For the next 60 kilometers, the highway travels south along a flat plain to Guerrero Negro providing panoramic views of flat-topped mesas to the east and the blue waters of the Pacific to the west. The prominent dark-topped mesas to the left are basalt overlying the lacustrine and lagoonal rocks to shallow water marine beds of the Paleocene Sepultura Formation. To the right, a volcanic hill forms the north end of Laguna Manuela.

The power company has erected poles with nesting platforms to draw the birds, mostly Osprey, away from the power poles. The osprey have built nests on some of them so be on the lookout for occupied nests.

The highway continues to traverse the relatively flat, gently undulating plain, occasionally dropping into and through numerous small arroyos as it crosses the expanse of Lanos del Berrendo referred to as the Antelope Plain. At the present time, there are only a few of the once abundant herds of antelope left.

The grasslands of the world are home to most of the great herds of herbivores. In Baja, the Pronghorn Antelope (*Antilocapra americana*) is the only large feral grazing ruminant that inhabits the vast Lanos del Berrendo. The herds of antelope have decreased due to hunting pressures (now illegal in Baja) and the destruction of the grasses the antelope feed on as this region was changed by agriculture and cattle grazing practices. Today, they have retreated to rougher, higher country unsuitable for agriculture or cattle grazing. Although they look like deer, the antelope are not members of the deer group nor do they have much connection with the Old World antelopes. The horns of the pronghorn antelope have a permanent bony core that is not discarded, although the outer horny, modified hair-like covering (keratinized skin) is shed once a year. Old World antelope never lose any part of their bony horns and they lack the outer horn sheaths.

108

80   The low dark hill to the right is a small basalt cone (*See* 4:98). The point of Guerrero Negro comes into view in the far distance.

The dune fields to the right parallel the Pacific coast and extend from the edge of Laguna Manuela to the north end of Scammon's Lagoon.

This **sand dune building** has resulted from the southward transportation of sand along the coast by longshore currents. The currents have caused a sand bar to build south of Punta Manuela. The winds pick up the sand and carry it into the lagoon and landward, away from the coast. This builds a set of migrating coastal dunes both on the bar and on the landward side of the lagoon. As the dunes progress inland, they move farther from the supply of coastal sand and soon become stabilized by plants that utilize the moisture trapped in the sand. The vegetation between these migrating dunes is quite sparse and consists of a few salt tolerant species (halophytes) such as Pickleweed and Sand Verbena. The tops of the well drained, less saline soils of stabilized dunes support specimens of Mesquite, Atriplex, and Ice Plant.

82.5   The highway passes Rancho San Angel and a low hill of basalt to the right. The arroyos adjacent to the highway are seasonally brightened by flowering Agave, Broom Baccharis, Tamarisk, and Barrel Cactus.

DATILILLO FENCE

LIVING FENCES: The fence posts in many other parts of Baja are often made of the cut branches and trunks of several cactus species, primarily Cardon, Pitaya Agria, Ocotillo, and Datilillo.

Numerous desert plants in Baja reproduce in this fashion. Several species of Cholla cactus have pads that readily fall from the main plant, develop adventitious roots, and become new plants. These plants are said to have propagated by vegetative asexual reproduction. Species that reproduce in this manner include Cardon, Yucca, and Datilillo. Vegetative reproduction allows desert plants to avoid the costly water and energy-consuming processes of sexual reproduction and permits the spread of plant species even when its too dry for seed production and the establishment of new plants by seed germination.

Most of Lanos del Berrendo is underlain by a thin layer of Pleistocene limestone deposited when this part of the peninsula was covered by a warm shallow sea. Limestone is alkaline and makes it harder for plants to establish themselves. As a result, the Lanos del Berrendo is quite sparsely vegetated by the Palmer's Frankenia-Datilillo plant community with scattered Cholla.

PALMER'S FRANKENIA-DATILILLO PLANT COMMUNITY

A **plant community** is a regional assemblage of interacting plant species characterized by the presence of one or more dominant species. Some plant communities are named for the tree or shrub species that are dominant in them. The term "dominant" refers to one or more plant species that may be the largest or most abundant species in a community. Because of the foliage cover or the extent of their root systems, dominants have a strong influence on the local ecology of the plant community.

83   To the left rear is a reddish cinder cone on the prominent dark mesa.

98   The dark peak at 3 o'clock is a basalt cone forming Punta Santo Domingo at the north end of Laguna Manuela. It has a shield-like shape with a prominent, steep, dark, eruptive cinder cone in the center. The gentle slopes are lava flows while the steep central slope is the cinder cone.

99   The road material quarries on the left side of the highway expose the marine Pleistocene limestone.

115   Almost nothing over a foot tall grows from here to Guerrero Negro. The low gray (10-30 cm) tufted mound-sloped plant that covers the Lanos del Berrendo is a bursage, Palmer's Frankenia (*Frankenia palmeri*).

PALMER'S FRANKENIA is a stiff woody shrub that grows on sandy areas, salt marshes, and alkali flats throughout the peninsula. The white flowers, which bloom in November, brighten the entire region around Guerrero Negro. A walk through the ground cover of Palmer's Frankenia reveals epiphytic foliose lichens and extensive algae mats that nearly cover all of the bare ground between the Bursage. Algae mats such as these are nitrogen fixers that add valuable nitrogen compounds to the desert soils.

A taller gray-green leafed introduced Australian weed commonly known as Australian Saltbush (*Atriplex sp.*) grows near the highway shoulders or on the sand dunes. This weed, now a naturalized native, was originally cultivated for cattle forage, but escaped and now grows extensively in saline soils and along the shoulders of the highway throughout the peninsula.

Saltbush is a halophyte that has modifications that enable it to grow in very salty dry soils. The seeds of the Saltbush contain chemical growth inhibitors that dissolve when exposed to sufficient amounts of water. These inhibitors ensure that the seeds will only germinate when sufficient quantities of water are available to allow the seedling to complete its life cycle.

Saltbush can grow in soils with a high salt content. The salt is absorbed into the plant through its roots. The leaves can be used as a flavoring for foods, and the parched ground seeds make a tasty coarse meal or fine flour.

118   As the highway continues southward, a 120-foot high metal eagle monument named Monumento del Águila dominates the highway. This monument was built on the 28th parallel to commemorate the completion of Highway 1 that joins the states of Baja California and Baja California Sur.

120   The highway approaches and crosses a semi-stabilized dune field.

123   The coastal lagoon of Laguna Manuela will be in view to the right of the highway for the next half-kilometer.

EAGLE MONUMENT

128 = 220.5 State Line eagle monument at the 28th parallel. The kilometer markings will now descend to zero points. The 28th parallel marks the boundary of the Pacific Time Zone and Mountain Time Zone.

OSPREY IN NEST

For years, there was a huge nest of sticks belonging to a pair of Ospreys nested on the sign on the east side of the monument. At some point it was removed; however, there are a number of nests in the area including out on the old loading dock west of town

216.5 The highway forks here. The left fork leads to San Ignacio; the right fork leads 2 kms through super-tidal flats, that are intermittently flooded by very high tides, into the town of Guerrero Negro. To see the birding and whales at the barge loading area, go into town, cross a channel, and turn right at the main cross street after the bend.

GUERRERO NEGRO SALT WORKS FROM AIR

THE WORLD'S LARGEST SALT PRODUCING REGION: Over 5,000 people live in the town of Guerrero Negro. The town's name "Black Warrior" refers to a whaling bark (ship) wrecked in the lagoon in 1858. According to Exportadora de Sal, the salt company that owns the town, the economy of Guerrero Negro is supported by the "world's largest" salt-producing operation.

To the south of the highway on the vast tidal flats of Laguna Ojo de Liebre (Scammon's Lagoon), are man-made salt-pans which cover over 800 square kilometers. Exportadora de Sal S.A. has diked some of the shallow parts of the tidal flats of Scammon's Lagoon, forming shallow evaporating ponds that are approximately 100 meters square and one meter deep when flooded with sea water. As the sun evaporates the sea water, the sulfates and carbonates precipitate out, leaving a brine solution that is moved to adjacent salt-pans to precipitate the halides (table salt, NaCl, is the most common). When the brine has completely evaporated, the salt precipitate is loaded into triple trailer bottom-dumping trucks (150 feet long), and transported to loading docks southwest of town. The salt is loaded on barges and lightered to Isla Cedros where it is shipped to the United States, Japan, Canada, and Mexico.

HOW SALT CAME TO MEAN SALARY: The word "salt" comes from the Latin word "sal." During the time of the Roman Empire, soldiers were paid partly with the coin of the empire and partly with salt. If a soldier failed to fully perform his duties, his ration of salt would be cut or withheld because he was "not worth his salt." The English term "salary" was derived from the custom of partially paying workers with salt. Today, salary refers to a fixed payment received at regular intervals for services rendered.

Many people visit the lagoons to watch and even touch the whales. If you are just passing through or running short on time there is a good whale viewing spot west of town. It is reached by driving out a long causeway through the marshes west of town. Take the first road to the right after passing the Pemex station, a canal, and making the big turn into town. This is also an excellent bird watching area. At low tide, the mudflats are covered with bay scallops.

THAR SHE BLOWS – THIS WHALE WAS WITHIN 100 METERS.

Fairly close **WHALE BREACH** - You have to be quick and lucky to catch them out of the water.

SALT MARSH AT HIGH TIDE

Low tide will expose **BAY SCALLOPS ON THE MUDFLAT**. There were millions of them. Dinner anyone? Yum! Yum!

BAY SCALLOPS

Log 5 - Guerrero Negro to Santa Rosalía [217 kms = 141 miles]

Guerrero Negro to Abreojos road. *The highway crosses the flat Vizcaino Plain, paralleling the Cretaceous syncline, past dune fields and cultivated fields as it approaches tilted Miocene volcanic mesas. Cretaceous metamorphic and sedimentary hills form the skyline to the west. As the highway approaches the Abreojos road, a series of steep-sided volcanic plugs are to the right.*
Abreojos Road to Tres Virgenes. *The highway turns east to cross the peninsula on the tilted surfaces of volcanic flows passing near numerous volcanic centers and dune fields. East of San Ignacio, a strato-volcano is approached and closely passed.*
Tres Virgenes to Santa Rosalía. *The highway drops down a series of fault controlled grades through volcanic rocks and then marine Pliocene rocks to the copper mining area of Santa Rosalía.*

217   To go straight to Santa Rosalía, take the left fork at the triangle.

215   For many kilometers, the highway will pass through a semi-stabilized dune field consisting of sand blown from Scammon's Lagoon. The stabilized sand dunes are vegetated with Datilillo, Mesquite, Jumping Cholla, Cardon, Palo Adan, Purple Bush, and Atriplex. The alkali flats between the dunes are sparsely dotted with stunted annuals (*See 3:21*).

STABILIZED SAND DUNE

208.1 The sign "Parque Natural de la Ballena Gris" marks the turnoff to Laguna Ojo de Liebre (Scammon's Lagoon) and a gray whale viewpoint 27 kilometers (17 miles) down a dirt road (the one perpendicular to the *Baja Highway*). The conditions of this road vary with the weather. (Check in town.)

Main view point - It is a long way to the lagoon and the road is sandy with deep dunes. The lagoon is wide and the whales are not close. We saw more whales; however, they were at greater distances than at the loading dock. Try one of the many whale watching guide outfits in the area. Expensive!

If you can secure a panga (for a fee) you can get out near the whales. If there is a stiff wind blowing (much of the time) the local Panga guide is not willing to chance taking you out. If their engine quit, you would be isolated and drifting a long way down lagoon. You cannot launch your own boat.

MAIN LAGOON VIEWPOINT - The whales were not even close.

THAR SHE BLOWS! The best known of the great whales and the one most often seen along the Pacific Coast of Alta and Baja California is the California Gray Whale (*Eschrichtius gibbosus*). Gray whales feed in summer (mid-May to mid-October) in the plankton-rich western Bering Sea and in the Chukchi and Beaufort Seas. In winter (mid-October to early December), they migrate south along the more shallow Pacific coastal waters of North America to the various bays and lagoons of Baja California, especially Laguna Ojo de Liebre (Scammon's Lagoon), where they breed and give birth to their calves (early December through February). In the spring (March and April), they travel north along the coast to arrive at their arctic summer feeding grounds about the middle of May. As they travel fairly close to shore, it is relatively easy to see migrating Gray Whales from high points of land or from boats. It is estimated that the whales migrate about 10,200 kilometers (6,250 miles) each way every year by traveling 60 to 80 nautical miles (69 to 92 miles) per day at a speed of 4 to 4.8 knots per hour for 15-20 hours a day. As they travel, they frequently raise their heads out of the water "spy-hopping" to get their bearings. It is believed that they find their way on the long trip by memory and vision.

THE REPRODUCTIVE CYCLE OF A CALIFORNIA GRAY WHALE

June to October	24 hrs. of daylight allow continuous summer feeding in Bering, Chukchi, and Beaufort Seas
November - December	Migration south, conception in route mid-Oct. to early December
January - February	Wintering in Baja's lagoons
March - April - May June to October	Migration north - March to mid-May Summer feeding in northern seas mid-May to mid-October
November - December	Southward migration
January - February	Birth of calves and nursing (7 months)
March - April - May	Northward migration, nursing continues
June to October	Summer feeding in northern seas Nursing ends in August

The California Gray Whale was once more numerous than it is today. In the 1800s Charles Scammon, a famous whaling captain, estimated the gray whale population to be 30,000. However, Scammon and other whalers slaughtered the gray whales by the hundreds for their valuable oil. By 1937, it was estimated that their numbers dropped to approximately 100. In 1938, they were given complete protection by an international treaty. Fairly recent counts of migrating whales from the shore and in the lagoons of Baja from the air have shown that the gray whale population has increased to approximately 15,000 to 17,000, and is still increasing yearly.

Gray whales are mysticeti or baleen whales that feed mainly on small crustaceans (primarily a red shrimp-like organism known as lobster krill) and plankton that are filtered from the water by sieve-like plates of "hair" (baleen) in their mouths. The captured organisms are "licked" off the baleen by the whale's tongue and swallowed. Most feeding occurs during the Arctic summer months (mid-May to mid-October), but some occurs during migration and in Baja's lagoons.

Baleen whales are thought to have evolved in the warm temperate waters of the western South Pacific during the Oligocene. During the late Cenozoic, they moved into the rest of the Pacific that included the northern seas presently inhabited by the California gray whale.

119

206.5 There are a few scattered Datilillo, Cardon, and Creosote Bush in this area. In early 1979, Pemex drilled an oil well approximately one kilometer east of the highway. In the future, oil and natural gas may be another important source of income for residents of this region.

OIL AND GAS: The geosynclinal structure of the Vizcaino region is identical to the great Valley of California. Geosynclines such as these produce much of the oil and gas in the world. Mapping in the Vizcaino and Magdalena Plain has indicated the presence of rocks and sedimentary environments favorable to oil and gas formation. After the traps where the petroleum accumulates have been delineated, this region may become a major oil producer.

202　Road material quarry in the Pleistocene limestones.

194　In this region of the Vizcaino Desert, the soil has become so dry that even the dunes are sparsely vegetated. The plants present along the highway are predominantly the xerophyte: Bursage.

PLANTS ARE PREDOMINANTLY THE XEROPHYTE: BURSAGE

189.5 Turnoff on Mexico 18 leads to El Arco, a gold mining town and the site of a porphyry copper deposit in the metavolcanic rocks. The old road went inland at Rosarito skirting the hills to El Arco. It then headed south close to Sierra San Francisco to join the highway near Abreojos Junction. Most early travelers used the rougher road to avoid the sand near the coast.

Porphyry copper deposits are usually mined in open pits on a large scale. The El Arco area promises to be one of the largest copper deposits in the world. They estimate 600 million tons of .7% copper ore.

187　The highly eroded volcanic mesas of the Sierra San Francisco lie ahead. To the right front are the volcanic plug domes of Sierra Santa Clara. To the right rear are the metamorphic and igneous rocks of Sierra El Placer. To the left and left rear are the metavolcanic hills near El Arco that are part of the southern end of the Peninsular Range Batholith. These mesas, domes, and ranges may not be visible at all on hazy days.

As you cross the Vizcaino Desert you will notice changes in the vegetation. However, most of the species are everywhere, they just change in relative abundance. Datilillo, Cardon, Palo Adan, Creosote, and Pitaya Agria all put on a show for one reason or another, locally becoming the dominant species.

172   For the next 7 kilometers, the sparse low vegetation of the flats is suddenly replaced by a very dense forest of Datilillo, Cardon, Palo Adan, Creosote, and Pitaya Agria. The undulations in the highway are dunal sand blown over the area.

165   The dense forest is suddenly replaced by the sparse low vegetation of Palmer's Frankenia with scattered Datilillo, Cardon, and Palo Adan.

162   Plants commonly seen adjacent to the pavements edge along this stretch of the highway include several opportunistic species such as the tall yellow-flowered large-leafed Indian Tree Tobacco, Cholla, and grass.

A major **transpeninsular fault** roughly parallels the road in this area and provides a groundwater barrier that the farming enterprises are mining for irrigation. This fault is easily seen in the seismic and magnetic geophysical data that has been collected for this part of Baja.

There is little Peninsular Ranges basement along the gulf south of this area. The highway was built parallel to the axis of a **Cretaceous syncline**, located a few kilometers to the west, with approximately 30,000+ feet of Jurassic to Paleocene sedimentary rocks. This geosyncline crosses the peninsula and

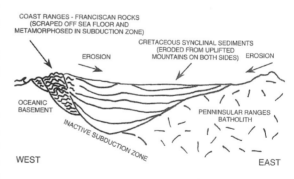

heads southeast toward Loreto. The southern part of the peninsula to the cape region is entirely underlain by the Cretaceous syncline and overlain by Tertiary sedimentary and volcanic rock. The Sierra San Andreas (similar to the Coast Ranges of California) with the Franciscan assemblages and extensive exposures of sedimentary rocks is to the right in the distance.

154.5 Ejido Francisco J. Mujica. The Sierra San Francisco to the left is a series of lava flows, folded into a broad arch, and tilted to the west. Some of the peaks that rise above the main arch are volcanoes.

152 Before Kilometer 151 Creosote Bush, Palo Verde, Cardon, Garambullo, Pitaya Dulce, and Pitaya Agria are once again more abundant. After Kilometer 151, Palo Adan and Datilillo along with many species of cactus become dominant.

144   Road to Bahía Tortugas. The agricultural cooperative and experiment station of Ejido Vizcaino is located at this junction.

123.7 Microondas de Los Ángeles is to the right near a low volcanic hill.

WHERE ARE THE ANIMALS? Besides birds, have you seen any other animals along the highway? The most common answer to this question is, "No." However, despite the harsh conditions of deserts (intense heat during the day, cold at night, and scarcity of water, vegetation, and cover) many kinds of animals do live successfully in this desert. The larger desert animals are shy, and many are nocturnal and unseen by daytime travelers. Those that are out during the day tend to be small and light in color for camouflage and protection against heat and dehydration. Smaller desert mammals are mainly fossorial (fitted for digging, living in burrows). The lower temperatures and higher humidity of burrows help to reduce water loss by evaporation. The large desert ungulates (hoofed animals) such as pronghorn antelope cannot escape the desert heat by living in burrows. The glossy pallid color of their fur reflects direct sunlight, and the fur itself is an excellent insulation that helps to keep heat out. Additionally, heat is lost by convection and conduction from the underside of the antelope where the pelage (fur) is very thin. Ungulates conserve water by eliminating concentrated urine and dry feces.

SIDE BLOTCHED LIZARD

The Side Blotched Lizard is common in Baja. It eats scorpions, spiders, mites, and ticks. It loves to bask in the sun on rocks.

To see animals in the desert, the best time to look is at night. However, during a walk into the desert during the day, you may see many birds and small inhabitants such as furry velvet ants (wingless wasps), lizards, black and red ants, and snakes. At the very least, you will see signs of desert inhabitants such as the **tail drags** and **burrows** of rodents; **tracks** of birds, rodents, lizards, rabbits, coyote; and **scat** (feces) of animals. Good field guides are available for identifying the tracks and scats of animals.

Some birds that you might expect to see in the Vizcaino Desert are:

COMMON NAME	LIKELY LOCATION
Amer. White Pelican	Flying over water
Cactus Wren	Among cacti
Calif. Brown Pelican	Flying over water
California Quail	Searching ground litter for food
California Thrasher	Searching ground litter for food
Gilded Woodpecker	Among cacti
Gray Thrasher	Looking in ground litter for food
Greater Roadrunner	Running on the ground
Ladder-back Woodpecker	Flitting through the air
LeConte's Thrasher	In sparse vegetation, flies when necessary
Red-Tailed Hawk	Soaring in skies
Sage Sparrow	Perching in desert, sage brush or chaparral
Turkey Vultures	Feeding on carrion or soaring
Western Meadowlark	On fence posts or wires

118.3 Dirt road into the Sierra San Francisco.

THE SIERRA SAN FRANCISCO to the northeast is eroded out of a series of volcanic rocks. These rugged mountains rise from the surrounding desert to more than 6,000 feet and stretch northward for 30 miles. The Sierra San Francisco was once a rather simple basalt plateau roughly circular in outline. It is now folded into a broad westward-tilted arch crowned by Pico Santa Monica (7,034 feet). Erosion has produced the rugged landscape seen today.

PREHISTORIC CAVE PAINTINGS: The mountains of southern Baja California from the Sierra San Borja south into the Sierra San Francisco and Sierra la Giganta are riddled with isolated caves painted by pre-historic people. These caves, located in some of the most inhospitable remote mountainous terrains of the peninsula, contain paintings of people, plants, unknown symbols, and animals. The figures are larger and more numerous than those of the famous prehistoric panels of Lascaux and Altamira.

Jesuit missionaries were probably the first "white men" to see Baja's painted caves. Leon Diguet, a French naturalist, was the first to publish an account of the caves after he visited them (1894). The caves were largely forgotten until Earl Stanley Gardner visited them in 1962 by helicopter. Since then, many have visited the caves. An excellent book with beautiful photographs entitled *The Cave Paintings of Baja California* has been prepared by Harry Crosby.

117 The only plants that seem to be growing here are Datilillo, Jumping Cholla, and Cardon. Upon closer examination there are Acacia, many low herbaceous annuals, Mallows, Brittlebushes, and Sand Verbena.

112 The two flat-topped buttes ahead are composed of Miocene sedimentary rocks capped by dark basalt. They are erosional remnants of the volcanic mesas of the Sierra San Francisco.

108 The highway leaves the very flat old lagoon surface and begins to climb the gently-sloping alluvial fans cut by washes. Common Ravens are often seen in this area.

COMMON RAVEN OR AMERICAN CROW?
There are three easy ways to distinguish between these two black birds: tail shape, habitat preference, and flight patterns. Color and body size are not good criteria to use as both birds are black and may vary in size depending on their age. The chart will help you identify these birds:

	Raven	Crow
Tail shape	Diamond	Fan
Habitat Preference	Deserts & cities throughout peninsula	Not usually seen south of San Quintín
Flight	Can glide for more than 3 seconds	Have a short gliding time - less than 3 sec.

102 Some of the Elephant Trees, Palo Adan, and Datilillo that grow here support the harmless epiphytic ball moss, *Tillandsia*.

98.5 The Pleistocene limestone that covers parts of the Vizcaino Plain is exposed along the highway. This limestone was deposited in a shallow warm sea that covered all of this area within the last million years. The lagoons of San Ignacio and Scammon are small remnants of this embayment that isolated the Sierra San Andreas and the Santa Clara Buttes as islands.

SANTA CLARA BUTTES

97.5 At the turnoff to Punta Abreojos, the gray Miocene marine sedimentary rocks of the San Ignacio Formation are exposed under the basalt on the high hill to the right. Just beyond the turnoff are more exposures of the Miocene sedimentary rocks. The Santa Clara Buttes to the right are Miocene (10-15 million years) andesite plugs that have intruded into and domed the Cretaceous and Paleocene sedimentary rocks. Their temporal relationships and geology are identical to the Marysville Buttes of the Sacramento Valley in California.

The isolation of the Vizcaino: The road to the right leads 50 miles west to an isolated section of the Pacific coast and the small cannery and fishing settlement of Abreojos. It is generally a fairly good graded road that most two-wheel drive vehicles can make. The area around Punta Abreojos is one of the favorite surfing spots on the peninsula. The break of the waves off the point enables some surfers to have runs up to half a mile.

97   Southeast of this point the vegetation is dense with Datilillo, Cardon, Jumping Cholla, Pitaya Dulce, Palo Blanco (first stands), Elephant Trees, Lomboy, Leatherplant, Palo Verde, Acacia, and occasional Ball Moss.

90   There is a large dip and bend in the highway. The flat-topped mesas ahead on the right are the mesas of the Sierra San Pedro.

88.2 Road to Microondas Abulón microwave tower. The view from this tower is sweeping. On a clear day you can see south to Sierra San Pedro and the mesas of Cuarenta, San Ignacio Lagoon, the Sierra Santa Clara plugs as well as the mesas of Santa Clara to the west. To the northeast is the mountain mass of Sierra San Francisco.

The quarry at this location is in the San Ignacio Formation and has yielded some spectacular agatized (a translucent cryptocrystalline variety of white

125

quartz) Turritella (*Turritella sp.*) fossils that are in a yellowish layer just below a light-colored volcanic tuff (ash) bed.

AGATIZED Turritellas: Approximately 11 million years ago this area was covered by a shallow sea. Molluscs including *Turritella ocoyana* flourishied in the sea. A volcanic eruption spewed ash over the area resulting in a mass mortality of these molluscs. Over the years, groundwater leached the calcite from the shells and replaced it with silica from the volcanic ash producing the agatized Turritellas.

AGATE QUARRY

86   The highway climbs and skirts to the south of the Sierra San Francisco. The road follows what amounts to a stripped-dip slope developed on a resistant bedding surface with basalt hills along the highway.

81   The highway sharply dips into an arroyo cut into the lava surface. The small red-brown conical hill to the left is an extrusive volcanic plug.

78.5 Several dark conical-shaped cinder cones dot the landscape. The very large dark mountain in the far distance ahead is Volcan las Tres Virgenes.

78   Turnoff to a paved jet airport (Areopista) that serves San Ignacio.

76   The road passes along the edge of the dark basalt mesa with views of the oasis of San Ignacio. These mesas are capped by dark basalts and underlain by the white and greenish San Ignacio Formation.

Mina (1957) defined the **San Ignacio Formation** for the light-gray sandstones, tuffaceous sandstones, and tuff outcropping near the town of San Ignacio. The San Ignacio Formation is a marine facies of the Comondú Formation that overlies and interfingers with the Isidro Formation. The San Ignacio Formation is Middle Miocene in age in the San Ignacio area; a volcanic tuff bed was K/Ar dated at 11 m.y. Some beds contain the agatized Turritellas (*Turritella ocoyana* and *Turritella inezana*). Further south in the Magdallena Plain of the same unit has been K/Ar dated at 22 m.y.

HUERTAS DE SAN IGNACIO - Huertas is the Spanish name for garden or orchard, and it is aptly applied to the desert oasis that surround the two small communities of San Ignacio and San Lino. It is reported that over 80,000 date palms grow here. Date Palms (*Phoenix dactylifera*) were introduced into mainland Mexico by Jesuit priests and have subsequently been cultivated in San Ignacio (since 1728), Mulegé, Loreto, and San José del Cabo.

74   A date of 11 m.y. was obtained on the San Ignacio Formation in this road cut.

RÍO SAN IGNACIO

73   This side road leads 3 kilometers to San Ignacio. Ground water surfaces in the arroyo above town to form a flowing river with an oasis of imported European date palms. It's well worth a turn off into the town to see the square that is shaded by large Indian Laurel Fig trees and the impressive stone Misión San Ignacio de Kadakaman built by the Dominicans in 1786. Hotel Mission de San Ignacio is located on the road into town.

SAN IGNACIO DE KADAKAMAN

ZÓCALO SHADED BY INDIAN LAUREL FIGS

69   The tuffaceous sandstones of the Comondú form the outcrops along the edges of the arroyo. The basalt is dated at 9.7 my. The highway begins an ascent of a grade bordered by a dense cardonal. The dominant plants in this area are Cardon, Cholla, Elephant Trees, Palo Verde, and Pitaya Agria.

67   Good view of the composite strato-volcano Volcan Las Tres Virgenes, composed of a combination of clinkery basalts and andesites and pyroclastic material in the form of lahars and associated breccias and mudflows.

Volcanoes are classified on the basis of the cone that they produce. The type of cone depends on the composition and viscosity of their lavas. Basaltic lavas with a lower content of silica (50%) flow easily, while rhyolitic lavas, which have a higher content of silica (70%), resist flowing (are more highly viscus) and are more explosive. Basaltic lavas from cinder cones have a higher content of gas and are more explosive. As the gases are released, the basalt becomes fluid and forms shield cones and lava plateaus.

The world's well known volcanoes are **Strato-Cones**; examples are Mount Vesuvius and Krakatoa. They are usually 2-3 miles high, tend to dominate the landscape, and are the most destructive types of volcanoes.

The table below relates the types of cones with activity, rock types formed, and material ejected.

TYPES	ROCK TYPE	CONE	MATERIAL
EXPLOSIVE	BASALT RHYOLITE	CINDER DOMES	CINDERS+ FEW FLOWS
INTERMEDIATE	ANDESITE	STRATO	TUFF & LAVA
QUIET	BASALT	SHIELD	LAVA FLOWS+ FEW CINDERS
FISSURE	BASALT	PLATEAU	THIN LAVA FLOWS

57   This straight stretch of road presents views of the volcanic mesas and plains of the middle part of the peninsula. Sierra San Pedro is to the right, and the relatively unexplored rugged volcanics of Sierra San Francisco to the left. The large dark mountain ahead is Volcan las Tres Virgenes.

VOLCAN LAS TRES VIRGENES: This active composite volcano is thought to have erupted as recently as the early half of the 18th century. There is a report from one of the early Spanish missionaries of an eruption in this area in 1746 that may well have been a description of the most recent eruption of Volcan las Tres Virgenes. There is another report of some smoke in 1857.

An **ACTIVE** volcano has a record of historic activity. The lack of active volcanoes in Baja may be due to the short recorded history of Baja or due to the fact that the sliding action of plates does not tend to produce volcanic action. Tres Virgenes is an active volcano.

129

A **DORMANT** volcano does not have a record of historic activity. It does however have a relatively good shape and shows signs of recent activity. Most of the Cascade Volcanoes such as Mount Rainier are considered dormant.

An **INACTIVE** volcano shows no sign of activity and shows significant cone erosion. A great example is Crater Lake in the state of Oregon.

51   The wind and the abundant supply of sand in this area have combined to form the dunes that cover the slopes of many of the hills.

43.5 Some of the tallest and most beautiful Elephant Trees (*Pachycormus discolor* and *Bursera microphylla*) of the peninsula grow on this lava flow.

The vegetation on the flat plain (Llano) between and around the lava flows consists of Brittlebush, Mallow, Palo Verde, Creosote Bush, Leatherplant, Pitaya Dulce, Cardon, Teddy Bear Cholla, Hedgehog Cactus, Garambullo, Jumping Cholla, Palo Blanco, Palo Adan, and Lomboy.

40.6 The first lava flow is one of the few places on the peninsula where two types of Elephant Trees (*Pachycormus discolor* and *Bursera microphylla*) occur together close to the road. A walk onto the flow leads into a jagged landscape with Elephant Trees. This is also a good place for a close-up view of the floral dominants of Llanos de Magdalena area.

Elephant Trees - represented by the unrelated species of the two genera *Pachycormus* and *Bursera*. They grow luxuriously as trees that store water in the cortical cells of their elephantine trunks. The *Bursera* are distinguished from the *Pachycormus* by the incense-like odor of crushed leaves or fruits.

PACHYCORMUS   **BURSERA**

LOMBOY

Lomboy is a member of the spurge family seen on the flats and mesas of this region. This plant produces an acrid sap and remains leafless most of the year. This adaptation reduces water loss by transpiration in the extremely arid environment of the Magdalena Plain desert. Leaves appear after each rain and turn a distinctive red before falling off as the soil dries out again.

Palo Verde is a green barked tree that produces legume-like "bean" pods. However, Palo Verde is a member of the Senna family. Like Lomboy, this tree is normally leafless. The green trunk is a cladode, a stem that acts like a leaf, that performs photosynthesis in the absence of leaves (*See* 9:210). The Palo Verde trees are often covered with lemon-yellow flowers in the spring.

Palo Adan is a relative of the Ocotillo. However, Palo Adan has thicker branches, a trunk, smaller flowers, and normally grows from Parallel 28° south to the Cape Region in the clay and granitic soils of alluvial plains. Ocotillo is not usually found below the 28th parallel and is replaced by its look-alike relative, Palo Adan (Adam's stick) *(See 4:16.5)*.

Creosote Bush is a very common resinous desert shrub of the southwestern U.S. and Baja. Unlike many other desert plants, it is green year-round. The waxy coating on its leaves reduces water loss by transpiration that allows the shrub to withstand long periods of extreme drought. The strong-smelling resinous leaves resemble the odor of the coal tar distillate creosote.

Other plants associated with these plants are Jumping Cholla, Cardon, and pavement edge opportunists such as Brittlebush and Indian Tree Tobacco.

131

PALO ADAN   **CREOSOTE**

TRES VÍRGENES

40   The highway passes close to the west side of Volcan las Tres Virgenes and its multiple lava flows. The ground for miles around is covered with pumice of an unknown origin that has been shown to be unrelated to the activities of Volcan las Tres Virgenes.

38.8   Rancho El Mezquital.

38.7   The highway parallels the edge of a blocky clinkery basalt lava flow from Volcan las Tres Virgenes. The deep clefts (cracks) on the flow were caused by the pressure of the molten interior that pushed on the solid crust much like the cracking on the crust of a loaf of bread. A number of unusually large Elephant Trees are growing on this lava flow.

38   The highway begins to climb through a small canyon on the western slope to the crest of the range and the Virgenes grade.

34.8 Top of Virgenes' grade where the highway begins a long descent into the Gulf coast drainage southeast of Tres Virgenes. Descending this grade used to be extremely difficult and dangerous. The very difficult switchbacks of the old road are visible over the scarp just beyond the crest of the grade.

CHANGING VEGETATION REGIONS: The highway descends into the eastern drainage of the Sierra San Pedro into the Gulf Coast Desert area of the Desert Region. The Gulf Coast Desert area extends south from Bahía de Los Ángeles along the eastern Gulf coast of the peninsula to just south of La Paz and includes most of the southern Gulf islands. The predominant plants of the Gulf Coast Desert are Cirio, Ironwood, Palo Adan, Creosote Bush, Brittlebush, and three species of Elephant Trees (*Pachycormus discolor*, *Bursera microphylla*, and *Bursera hindsiana*).

32.3 The highway crosses an eroded fault scarp that forms the base of the grade. This fault has a major lateral motion component. This fault and one to the north of Tres Virgenes may be causing a small rift area that is responsible for Tres Virgenes and two other volcanoes on the area. The hill to the left, Punta Arena, is a volcanic eruptive center that has domed the strata of Mesa El Yaqui into a hill along part of the Gulf Fault Zone.

TRES VIRGINES FROM THE EAST SIDE

31.5 Rancho Las Virgenes. The road to the left leads to a large geothermal project (Proyecto Geothermico Las 3 Virgenes) located a few kilometers northward. This project is under the control of the Commision Federal de Electricidad and will produce electricity by harnessing geothermal heat to heat water to steam, that will in turn drive a steam turbine generator to produce electricity for this part of Baja. They estimate a production of 15 MW.

30   The highway follows a relatively narrow surface developed on the flat-lying volcanic and fluviatile sedimentary rocks of the Comondú Formation.

18.5 This is the first view of the Sea of Cortez (on a clear day). The volcanic cone visible offshore is Isla Tortugas. If you take a short hike to the right, there is a good viewpoint near the top of the grade.

17.5 The highway begins a twisting descent of the infamous Cuesta (grade) del Infierno in the Barranca de las (Palm Canyon) through yellow Pliocene marine sedimentary rocks. It's easy to imagine this descent on the old narrow rutted dirt road, and to understand why it is infamous to those who made that descent before the road was widened, leveled, and paved. This section of the highway is very steep and has few turnouts or shoulders. On one of our photo trips, we were taking a photo at the first turnout when a large truck rolled over on the curve below us.

PLIOCENE SEDIMENTS FROM VIEWPOINT

THE THREE FORMATIONS OF THE SANTA ROSALÍA PLIOCENE: The Pliocene of the Santa Rosalía area has been divided into three formations separated by unconformities and characterized by faunas that are believed to be Early, Middle, and Late Pliocene (Wilson, 1948, Wilson & Rocha, 1957).

The Pliocene sedimentary rocks of the area were deposited near a shoreline in a shallow marine environment interrupted by the deposition of tuffaceous material from explosive volcanic eruptions. It contains gypsum deposits.

During the Pliocene, the fault along the Gulf was active and periodically uplifted the area west of Santa Rosalía. This uplift resulted in the deposition of conglomerates into the finer-grained sandstones and siltstones. Mineralizing solutions rose along the faults flowed through the more porous and permeable conglomerates and mineralized the clays to form the Boleo copper deposits.

The Lower Pliocene **Boleo Formation** is a succession of sandstones, tuffs, and conglomerates that contain copper and manganese deposits. Recently, the Cinta Colorado Tuff has been age dated at 6.8 m.y. This places this unit in the uppermost Miocene. The Middle Pliocene **Gloria Formation**, later renamed the Tirabuzon Formation, is a sequence of fossiliferous marine sandstone, siltstone, and conglomerate. It can be seen between Km 4 and 8 north of Santa Rosalía. The Upper Pliocene **Infierno Formation** is a succession of fossiliferous marine sandstones and conglomerates.

ACTIVE COPPER MINE

Atypical Ore Deposits: The Boleo deposits are thin gently-dipping tabular bodies of impervious clayey tuffs that overlie sandstone conglomerates. The main gangue mineral is a montmorillinite or bentonite clay that is not typical of copper deposits. Low temperature hydrothermal solutions appear to have ascended through cracks and faults in the volcanics. They were blocked by the impermeable tuffs and permeated the tuffs by diffusion. This left the copper minerals finely dispersed in the clay matrix

PALO BLANCO

RAILROAD RUNNING STOCK

GYPSUM CRYSTAL

CHURCH BY EIFFEL - AN EIFFEL THAT IS NOT IN PARIS

AN EIFFEL THAT IS NOT IN PARIS – This sheet metal church was designed by A. Gustav Eiffel, architect of the Eiffel tower in Paris. The prefabricated metal church was displayed at the 1898 World's Fair in Paris, France. It was stored for years, then purchased by Compagnie Boleo and brought to Santa Rosalía. Look inside for a view of Eiffel's structural genius.

0   The rustic French architecture and Eiffel's metal church are worth seeing. Turn right at the church and go up the hill to the mining museum, old train engines and rolling stock, and the Hotel Francis.

Log 6 - Santa Rosalía to Loreto [196 kms = 122 miles]

Santa Rosalía to Mulegé*.* *The highway follows alluvial fans along the narrow fault controlled gulf coastal plain fringed by Miocene volcanic mesas. Near* Mulegé, *it climbs the Miocene volcanic hills to descend into* Mulegé *with its Pleistocene terraces.*

Mulegé to Loreto*.* *The highway continues south along a major fault zone on the same coastal plains to* Concepción *Bay with the Miocene volcanic mesas on the right. The opposite side of the bay consists of faulted Miocene volcanic rocks. South of* Concepción *Bay, the road climbs through a pass in the volcanics and follows a major alluviated graben along the Gulf fault zone with the volcanic mesas to the west and faulted volcanic rocks to the east. Near Loreto, it passes through faulted marine Pliocene sedimentary rocks.*

ANOMURAN CRABS ON BEACH AND HARBOR

197   The beach south of the harbor has been covered by a blanket of Decapods (Anomuran crabs) which are related to hermit crabs. These Decapods wash ashore by waves and can measure at least a foot deep. (*See* Kerstitch & Bertsch - Sea of Cortez Marine Invertebrates).

Dominican Louis de Sales wrote in the late 1760s that "while the ship Venus was thus following the coast to enter the Gulf of Cortez, the lookouts noticed red patches of whales blood. The presence of the whale-ships within sight might justify this guesswork supposition which however, was soon contradicted by the reality for soon, as we advanced, we crossed over these patches and recognized that their red color was caused by a multitude of small, vermilion colored crustaceans. These crustacea were [like] big shrimps, but they had what shrimps lack, pincers like those of lobsters." The Gulf of California is referred to as the "**Vermillion Sea**" for good reason.

For the next two kilometers, the road passes exposures of the Pliocene Tirabuzon and Boleo Formations, and pink tuffs of the Comondú Formation.

193.8 The beds ahead are gypsum and tuffaceous sandstones of the Pliocene Boleo Formation.

191.5 The highway crosses Arroyo de Santa Agueda and the Santa Agueda Fault, and passes through exposures of both the Boleo and Infierno Formations to climb a grade through sandstones of the Pliocene Infierno Formation. The head frame of the abandoned San Luciano copper mine is below in the canyon to the right. The remnants of the town site of San Luciano (streets, foundations, etc.) are still visible on the gentle slopes behind the mining equipment. The hills around the mining region of Santa Rosalía are riddled with mines, shafts, and the remains of mining structures.

188.5 Ramal leads to Santa Agueda and the supply of water for Santa Rosalía. For several kilometers, there are good views of Isla San Marcos.

ISLA SAN MARCOS is a Miocene volcanic island covered with Pliocene marine sedimentary rocks that are composed of gypsum precipitates deposited during the Pliocene in shallow marine basins with restricted circulation in the Gulf. The barren whitish slopes near the south end of the

140

island are the gypsum deposits in the Pliocene Marquer Formation. The gypsum is mined as a retarder in Portland cement, a flux in copper smelting at Santa Rosalía, and in making Plaster of Paris. Some people see the dog Snoopy lying in wait when they look at the outline of Isla San Marcos.

The vegetation is Palo Verde, Elephant Tree, Cacti, Cardon, Cholla, Leatherplant, Lomboy, and many desert annuals.

SANDY SPITS AND BEACHES: Most of the features along Baja's shoreline are produced by erosion or deposition. Depositional features in this area are built of eroded and weathered materials brought down to the shore by streams from the eastern escarpment of the Sierra la Giganta or material eroded from headlands by waves. At Santa Agueda, muddy-sand has been transported south by the waves forming the muddy-sand spit of Santa Agueda, a fingerlike extension of the beach. Other materials have been moved by currents and deposited to form the muddy-sand beach of San Lucas Cove.

181   Palm-shaded cove of San Lucas. Vultures frequent this area and are often seen sunning themselves in the morning sun.

VULTURE SUNNING – Vultures are not afraid if you move slowly. This one is facing the camera with his back to the sun and I am 15 feet away.

179.2   Cobblestone-paved side road leads to Microondas San Lucas. There is a view of Volcan Tres Virgenes, the Santa Rosalía area to the northwest, and Bahía de la Concepción to the southeast. There are many Elephant Trees near the top.

168.2   This road leads to San José de Magdalena. The village, in a palm oasis, was a Jesuit visiting station of Misión Santa Rosalía de Mulegé. Padre Ugarte built a 500-ton ship for exploring the Gulf from Oaks cut in the area.

163   To the southwest is the high Gulf escarpment which is not very well defined in this area. Between Santa Rosalía and Mulegé, the Gulf Fault Zone is a broad series of faults that cause the rugged foothills west of the highway.

159   The flora along this section of the highway is characteristic of the Gulf Coast Desert and is dominated by Cardon, Palo Blanco, Desert Mallow (apricot flowered), cacti, and yellow-flowered annual composites. The annual composites and Desert Mallow are primarily growing opportunistically along the edge of the pavement or in the disturbed soils that border the highway.

155.5   The graded road to the left leads 20 kms. to Punta Chivato. Exposures of fossiliferous marine Pliocene beds can be seen in the sea cliffs on Punta Chivato.

145   The highway follows an alluviated plain before it climbs a grade at Kilometer 142 through the Comondú volcanics. This plain drains southwest into a gap in the hills to join Río Santa Rosalía in Mulegé.

139.5   Ahead and to the left is a multicolored hypabyssal plug that is one of several 20 million-year-old intrusive plugs exposed in this area.

137.7   This is the first view of the desert palm oasis village of Mulegé. The highway begins a descent through volcanic rocks.

Mulegé is a refreshingly beautiful spot on the banks of the Río Santa Rosalía de Mulegé (Rio Mulegé). It is located in a valley bottom whose sides resemble an extremely dry desert cactus garden. Mulegé began as a Jesuit Misión in 1705. Because of the presence of the river, an oasis of European date palms, citrus, olives, grapes, and oranges has developed.

DRYING DATES

135   Purple andesite breccias are located at the north end of the Puente Mulegé. The road to the left leads into Mulegé and along the north side of the river to El Sombrito.

MULEGÉ ESTUARY EL SOMBRERITO

MULEGÉ PRISON

On the bridge, there is a view of the Río Mulegé, the estuary, and the closed federal prison on the hill to the north (large white mission-like structure).

The conical hill behind the prison is the Miocene intrusive volcanic plug, seen as the highway entered from the north of town.

133.6   The road cuts contain altered gypsiferous Pliocene strata cut by

143

basalt dikes. Most of the yellow sandstone and siltstone layers in this part of Baja are probably Pliocene.

EL SOMBRERITO FROM LA SERINIDAD

PANGAS AT LA SERINIDAD – FISHING ANYONE?

131.5   Just before the entrance to La Serenidad Lodge, the road cut exposes yellow-brown sedimentary rocks of Pliocene age that contain fossil oysters. Overlying them are fossiliferous Pleistocene deposits of the Mulegé Terrace. The Mulegé Terrace is a 10-meter marine terrace of Late Pleistocene age that is extensively exposed in and near the town of Mulegé.

A STABLE SHORELINE: In this area, the Gulf shoreline has remained relatively stable since the Pliocene. There are nearshore Pliocene rocks exposed along the Mulegé Estuary and nearby coastal areas. These Pliocene rocks are overlain by the Pleistocene Mulegé Terrace with its diverse shoreline fauna. The present shoreline is only a few feet lower.

THE MULEGÉ TERRACE has been dated at 125,000 to 145,000 years (Ashby, Ku & Minch, 1987) and probably represents the Sagamonian 5e high stand of sea level. The fauna contained in the terrace deposits indicate a variety of environments including open rocky shoreline, protected sandy shoreline, and estuarine (Ashby & Minch, 1987).

126   To the left is a good view of Punto Gallito with its Pleistocene marine terrace and dune field. The north end of Punta Concepción comes into view.

124.5   A cinder blanket is visible in this road cut. The road to Microondas Tiburones circles around an ancient cinder cone. From the top of the cinder cone, there is a view of nearby Punto Gallito and Mulegé to the north. The alignment of three nearby hills (Punto Gallito and two unnamed hills to the south) corresponds to the trace of a fault. To the south, the length of Bahía Concepción and the entire Concepción Peninsula including the Gulf Fault Zone are visible.

BAHÍA CONCEPCIÓN: Twenty-five miles long, it is one of the most beautiful bays on the Gulf coast. Turquoise and aquamarine waters wash its steep rugged shores with the milky blueness of tropical seas the world over. The color of the waters, produced by colloidal limes derived from the shells and skeletons of a myriad of marine animals, is the same as the color produced by colloidal suspensions in travertine springs, such as Havasupai in Arizona. Biological, "shell-sand" beaches are found in sheltered coves, such as Santispac, Coyote, and El Requesón. Many of the coves are bordered by small mangrove swamps. Mountains covered with desert vegetation almost surround Bahía Concepción and make it a land of extreme contrasts between mountain, desert, beaches, and tropical blue water. One of the densest cardonals in Baja is at the south end of the bay.

For the next several kilometers, there are excellent, scenic views of Bahía Concepción and the Concepción Peninsula. Patches of cinders are exposed in the hills. These cinders are often utilized locally for road material.

123   Approaching Bahía Concepción, the sparce flora consists of Palo Verde, Cardon, Garambullo, Mesquite, Busera, and Creosote Bush.

115.8   As the highway crests over a low pass, the parking area to the left provides an excellent view of Bahía Santispac, one of the most beautiful little bays along this coast. The light, milky green patch in the middle of the bay is

145

due to shallow waters that barely cover a sandy shoal. The white sands in this area are not lithic sand, but materials derived from the abundant calcareous fragments of marine invertebrates.

BAHÍA SANTISPAC IN THE EARLY MORNING LIGHT

BAHÍA SANTISPAC AND MANGROVE SWAMP

WHERE ARE THE SANDY BEACHES? Granitic rocks are coarse-grained and, as they weather, yield the sand-sized grains that compose sandy beaches. Because there are very few granitic rocks or granitic-derived sediments in northern Baja California Sur, there are few lithic sand beaches along the Gulf between Santa Rosalía and La Paz.

The volcanic sediments, derived from the Comondú Formation which are prevalent in the southern part of the peninsula, are fine-grained. When they weather, they yield smaller mud or silt-sized grains. In areas of volcanic sediments, there are no real sand beaches. Hawaii is a good example of this phenomena. Most of the "sandy" beaches in this part of Baja are composed of bits and pieces of weathered shells and coralline material instead of granitic sand. This type of beach is called a biological beach.

BIOLOGICAL BEACH: The quartz and feldspar that form the major portion of most sands are too fine-grained in the volcanic rocks to form sand. The majority of these "sandy" beaches are composed of mollusc shells, algae, and coral skeletons. These "sandy" beaches originated from biologically derived calcium carbonate and calcium phosphate particles that have been weathered to the same detrital particle size as quartz sand particles. Since the "sand" of these beaches were derived from the shells of once living molluscs and corals, they are said to be "biological sands." However, there are a few beaches in central Baja that are formed of quartz sand particles.

A fault is exposed in the road cut just before the turnoff to Bahía Santispac. This fault roughly parallels the axis of the bay and may have been involved in the origin of the bay.

113.8   This side road leads to Bahía Santispac. This is one of Baja's most scenic beaches with its white biological sands and azure waters. The small red mangrove swamp on the west side of the beach is the northernmost one that is directly visible along the Baja Highway. Bird watching can be rewarding; pelicans, cormorants, and frigate birds are commonly seen here.

The highway will pass numerous swamps located in the tidal creeks and estuaries that dot the Gulf coast from Mulegé to the tip of the peninsula. The predominant plant of the swamp is the Red Mangrove (*Rhizophora mangle*), an evergreen shrub that forms dense thickets in or near salt water or on alkaline soils.

Soil and debris carried by the tides and stream runoff from the adjacent coastal environments are washed into the network of stilt-like prop roots of the red mangrove and become caught. Over the span of many years, this process slowly extends the shoreline of the peninsula into the Gulf. When fertilized, the white or cream-colored flowers of the red mangrove produce a brown fruit that matures into a plantlet before falling off the parent. The plantlet has a long tap root, and when it drops off the parent plant, it falls like an arrow into the mud and plants itself. Several other plants commonly found in or near these swamps are Black Mangrove, White Mangrove, Pickleweed, Sand Verbena, and several species of native palms.

147

111.5   Posada Concepción on Bahía Tordillo.

107   The beach below and the beach of El Coyote are both biological beaches. The highway continues to pass a number of these beaches that are signed as "Playas Publicas" (public beaches) where anyone can camp. A fee per car per night may be collected for "maintenance."

DRIVE IN THE GULF! Just south of El Coyote, you can drive down to a small beach that was once located along the old road. When earlier travelers drove down this part of the peninsula they found that the road took the route of least resistance, skirting the shoreline in most places and climbing into the hills when absolutely necessary. Below the new highway, the old road was forced to skirt a steep sea cliff. There were large boulders on the outside edge of the road about 13 feet from the base of the cliff. At low tide, the highway was awash and vehicles drove through 6 to 10 inches of water. At high tide, the road and the large boulders were often covered by water.

MAIN ROAD IN GULF

The vegetation on the slopes to the inland side of the highway appears to be sparse and dead because it matches the color of the metavolcanic soils. However, a closer look reveals that the vegetation is growing quite well and are composed primarily of Cardon, Palo Verde, and Elephant Tree.

COVE ON CONCEPCIÓN BAY

106.5   The highway turns inland to bypass a difficult section of the coast.

102 to 98   The highway passes through a dense Cardonal with Elephant Tree, Palo Verde, and several species of cactus.

99   Basalts form numerous barren talus slopes on the hills ahead. Talus slopes are a type of landslide that form when rocks fall a few at a time and form a slope of debris that accumulates over a long period of time.

TÓMBOLO OF EL REQUESÓN AT LOW TIDE

92   View of the tómbolo of El Requesón, Isla El Requesón, and a red mangrove swamp. Herons and egrets are often seen along the edge of this

149

swamp. Good clamming and snorkeling from the island have been reported. It's worth a visit.

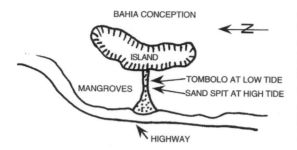

A **TÓMBOLO** is a sand bar that has been deposited by currents and/or wave action between a mainland and an island or between two islands connecting the two landforms. At El Requesón, a sand bar links the peninsula with Isla El Requesón at low tide.

SOPHOCLES IN THE 5TH CENTURY BC WROTE: *"One must wait until the evening to see how splendid the day has been."* With this sunrise at La Requison we did not have to wait.

90   The view ahead is of Playa Armenta, the last of the biological sand beaches that have dotted the shoreline since Mulegé.

89.5   To the left, the old road is visible immediately below the new highway. This is another reminder of the challenges that traveling the old road presented. As Baja travelers made their way along this rough road, one might all of the sudden smell pickles or mustard in the car. Upon further investigation, it would be discovered that one or more jar lids had come loose from all the jostling. The knowledgeable travelers stopped to check before all was lost or ruined. Due to the constant bouncing back and forth, labels would often wear off of the cans. Mystery pot luck anyone?

89   View to the left of Bahía Concepción. There are numerous, large alluvial fans on the Concepción Peninsula on the far side of the bay. These large alluvial fans were formed by streams that carved material from the hills of the Concepción Peninsula. There are no large fans on this side of the bay. This combination of numerous large fans on one side and a few small fans on the other is good evidence for the recency of movement on the Gulf Fault Zone that parallels the western side of the bay. The movement resulted in a down-drop of the western side while the eastern shore was uplifted. The large fans on the east side had a much longer time to form.

82.7   The cinders in this road cut are part of a Miocene cinder cone, one of several exposed along the highway over the preceding five kilometers. These cinder cones are probably distributed along a branch of the Gulf Fault Zone.

75.4   Large Cardons can be seen growing in a dense cardonal on the well-drained inland slopes. A quick look shows that they become smaller and sparser toward the shoreline of the bay; this is due to the more saline soils and poor drainage. Cardons grow best on well-drained slopes. Dense stands of *Bursera*, Copal, and Cholla are associated with the Cardon.

CARDONAL AND PLIOCENE BEDS

73.5   The flat-lying, whitish to pinkish rocks on the left in the far distance in a gap in the hills are marine Pliocene strata consisting of limestones, tuffs, and redbeds. They were formed in association with hot spring deposits.

71.5   As the highway ascends a grade, the view to the rear is the last of Bahía Concepción and the Concepción Peninsula.

70.8   The highway descends through road cuts that expose the andesite mega-breccias of the Comondú Group.

68.1   A large dark basalt dike cuts the light pinkish mega-breccia.

The vista opens onto the Loreto Graben. The alluviated valley that the road follows is a graben bounded on the right by the high eastern escarpment of the Sierra de Giganta and a horst composed of volcanic rocks to the left. The highway will continue to follow the axis of this graben to Loreto.

SIERRA GIGANTA FROM PASS

61.8   The highway passes through the palm oasis of Rosarito.

59.1   The graded road to the right leads to La Purisima, and Comondú.

56.2   The low hill 0.5 kilometers to the right is cut by two prominent nearly vertical dikes which are offset in a right lateral sense by one of the numerous traces of the Gulf Fault Zone.

55.1   The rancho on the left is Ascencion. Another road to the right leads to the Pacific side of the peninsula via Canipole (8 kms.), Comondú (76 kms.), and La Purisima (74 kms.). This is as far south as most people drove before the road was paved as the old road continuing on to Loreto was a dead end and used only by people going to Loreto.

53   The prominent white hill approximately one kilometer west of the highway is a hypabyssal intrusive plug along the Gulf Fault Zone.

52.5   El Bombedor, another small palm oasis, is a small roadside Rancho typical of this region.

36   There are ranchos on the alluvial plain. Most of these ranchos are located close to linear fault features clearly marked by lines of water-loving, more mesic vegetation. The faults in this region create zones of "crushed" rock that allow ground water to come to the surface. Between Kilometers 36 and 32, the hills to the left contain numerous north-south fault traces. These faults are part of the Gulf Fault Zone.

SIERRA GIGANTA AND A FAULT SCARP

35   Directly west, the high peak of the range is Cerro la Giganta, elevation 5,892 feet. A fault scarp offsets the fan directly below the peak. The fault parallels the highway halfway down the fan with approximately 10 m. of offset marked by a low line of hills that run through the fan (look at the line at the top of the tallest Cardon in the photo). Farther to the left are a series of small, sharp-crested hills. These are granodiorites dated at 145 and 87 m.y.

A MYSTICAL LAND CALLED CALAFIA: As the Spanish explored Mexico, they brought with them a book filled with legends of a mystical land called Calafia, an island somewhere to the west of Spain. Throughout their

153

explorations, they looked for the land of Calafia which was occupied by large "Amazon" women known as Gigantis. They were said to have lots of gold and jewels. When the Spanish traveled from Mainland Mexico across the Gulf to Baja California, they thought they had found Calafia. It was not until later that it was proven to be a peninsula. It is quite possible that California was named for the mystical island of Calafia, not for the Mexican word, Californax ("hot furnace"). When the Spanish settled in this area, they imagined they could see one of the Gigantis reclining in the Sierra de la Giganta. Can you see her?

29.8   This turnoff leads to the small date palm community of San Juan Bautista. To the southeast is the Boca San Bruno. Early Jesuits attempted to establish two visiting stations for the Loreto Mission in this area: Misión Guadalupe de San Bruno, 1683, and Misión San Juan Bautista Londó, 1687. Both attempts were unsuccessful.

29.5   The highway climbs through the first of the marine Miocene-Pliocene sandstones, siltstones, and conglomerates of the Loreto Embayment. *Some of the kilometer marks may be off between here and Loreto.*

A MARINE TECTONIC BASIN: Major movement of the Gulf Fault Zone appears to have initiated the formation of the Loreto embayment of Miocene(?)-Pliocene age; it is a thick sequence of conglomerate, sandstone, mudstone and limestone with pecten and oyster reefs, tuff, andesite, and basalt. Tuff beds and a lahar indicate volcanic action during the filling of this basin. At least one major angular unconformity and the lensing of conglomerate beds suggests intermittent deformation. Basalt (6.7 million years) intrudes and overlies part of the section that may be older than the rest of the basin. Other dates of 1.9, 2.1, and 3.3 m.y. have been obtained from tuffs interlayered in the marine sequence (Mclean, 1988, 1989). The syncline of the Loreto embayment is folded into a N-S anticlinal structure with numerous N-S and NW-SE trending faults that may have three to five thousand feet of cumulative displacement. The conglomerates contain granitic and metavolcanic clasts from outcrops to the west.

The Loreto embayment is one of the few tectonic basins in the southern part of the peninsula. The darker brown flat-lying Miocene rocks that cap the mesas to the west are as young as 10 million years old. This embayment was formed after faulting warped and uplifted the mesas. One dike, dated at 6.7 million years, cuts the sedimentary rocks. At this time, the northern part of the Gulf of California had already formed, and the more southerly parts of the gulf were being formed. The Loreto embayment was warped downward and received sediments from the surrounding highlands. The conglomerates indicate intermittent uplift of the same areas and more powerful streams. The yellow beds indicate wearing down of the source areas with less powerful streams and only finer material is carried far into the basin. The area was continuously faulted. Several tuff beds are exposed in the section that indicate continuing volcanism. These rocks are highly fossiliferous.

27.3   As the highway crosses a large arroyo, an auto-brecciated, andesite lahar is exposed on both sides of the highway.

26.5   Tuff is being quarried for use as building material from a white tuffaceous sandstone outcrop on the hill above the highway.

26   An oyster shell reef is exposed for a kilometer along the highway.

25.3   The oyster-shell reef and white tuffaceous sandstone parallel the highway for a short distance before crossing under the highway to the right to interrupt the yellow sedimentary beds exposed there. The surface of the hill on the right is veneered by oysters. The tuff yielded a 2.1 m.y. K/Ar date.

PLIOCENE MARINE SEDIMENTS

22.9   The yellow beds are well exposed in the hills near a small rancho along the highway. They represent a time of low sedimentation. Benthonic foraminifera from these beds indicate that they were deposited at shelf-slope break depths. The exposed conglomerates represent periods of uplift of the adjacent highlands along the faults in the Sierra la Giganta.

20.1   Dirt road to San Juan Londó. To the rear is a view of the Gulf. Just south of the water is a small, rounded volcanic hill. This was the site of a small visiting station of the Loreto mission. There are approximately 1,000 feet of conglomerates and sandstones between the yellow beds in this area.

19   To the southeast is the southern part of the Loreto embayment. The volcanic hills to the left are overlain by white tuffaceous sandstones and gray conglomerates of the lower part of the Carmen-Marquer Formation undifferentiated. These beds are thousands of feet thick. Common fossils include whale bones, *Argopecten sp.*, *Ostrea titan*, *Turritella sp.*, *Chama sp.*, *and Chione sp.* The southern part of the Loreto embayment was warped before the deposition of the upper beds that are also folded.

18.5   A pecten reef outcrops to the right above the highway. There is a view to the left into a distant amphitheater of yellow beds overlain by conglomerates with an isolated thumb of conglomerate in the center. The yellow beds have yielded very large oysters and whale remains.

PLIOCENE MARINE SEDIMENTS LORETO BEDS - The grey layers represent periodic uplift and erosion of the adjacent highlands while the yellow layers represent periods of little uplift and erosion.

13.8   The highway descends into a small arroyo. The limestone on the ridge to the left is much younger than and unconformably rests on the tilted edges of the Loreto sediments. There is a pecten reef in the road cut.

12.8   The highway crests a small grade and begins to follow Arroyo de Arce. The white volcanic tuff bed seen in the road cut is repeated five times for a cumulative separation of 600 feet on this bed alone. Sand dollars and pectens have been collected from the layer above the tuff bed.

FOSSIL SAND DOLLAR

12   Blocks of white tuff were mined at Rancho Las Piedras Rodadas to the right. This tuff is 3.3 million years old.

10.7   The highway crosses the crest of an anticline. The conglomerates and sandstones now dip to the south.

9.3   Side road leads to Microondas Loreto which is on volcanic rocks that yielded a 14.9 m.y. K/Ar (potassium/argon) date.

9   The road cuts reveal numerous small faults and scattered pecten fossils. The highway crosses the deep Arroyo de Gua.

A common bird in this area is the large spotted-breasted Cactus Wren. The **Cactus Wren** is the largest wren on the Baja peninsula, commonly seen feeding and nesting among thickets of thorny scrub, large cacti, or mesquite. They are identified by their streaked back, white eye stripe, and heavily spotted breast. Their nests are tucked into cholla or thorny bushes and they search through the ground litter for small invertebrates, usually insects.

MISIÓN NUESTRA SEÑORA DE LORETO

LORETO is Baja's first city. In 1697, Padre Juan María Salvatierra along with six soldiers landed at Loreto and established Misión Nuestra Senora de Loreto at the delta of a large perennial stream, Arroyo de Las Parras. Flood waters raging down Arroyo Las Parras have destroyed the town on 11 separate occasions since 1697. The town has also been damaged by earthquakes. A particularly large one occurred in 1877. Notice that the clock is not yet in the tower.

Log 7 - Loreto to Constitución [145 kms = 90 miles]

The highway continues south on the narrow Gulf coastal plain between the rugged Miocene volcanic mesas and the Gulf of California with its many faulted volcanic islands. South of Ligui, the highway climbs a grade through the Miocene volcanics and the Gulf fault zone to the mesa tops. It then follows a gentle canyon in the mesas to the flat Magdalena Plain.

120   **LORETO TURNOFF:** The uplifted fault scarp cliffs of flat-layered volcanic tuffs and sedimentary rocks to the west are the main Sierra la Giganta. From this point, it is hard to realize that there is a relatively flat, gently sloping tableland on top that extends west to the Pacific Ocean. The prominent peak to the right of the large pass in the Sierras is Cerro Pelon de Las Parras, a shallow intrusive andesite plug that yielded a 19.4 m.y. K/Ar date.

The low hills to the northwest are composed of 94 m.y. old metavolcanic rock and 143 to 87 m.y. old tonalite with a moderate to complex structure intruded by a dike system. The presence of older granitic and metavolcanic rocks at this point in the Gulf of California plus the lack of Peninsular Range basement in south-central Baja confirms geophysical evidence for the continuation of the Cretaceous syncline under most of Baja California, Sur.

CATTLE ON SAN JAVIER ROAD

South of Loreto, the Gulf Fault Zone and the highway converge until the highway is forced to follow a narrow hilly coastal plain between the Gulf and the steep cliffs of the Sierra la Giganta.

117.6   The road to the right leads 30 kilometers up Arroyo Las Parras to Misión San Francisco Javier de Vigge. This mission was founded in 1699 and was the second Jesuit mission on the peninsula. It has the distinction of being the only original mission church that has remained intact.

The road crosses a Comondú section that is different from the other sections crossed by the main highway at San Ignacio and south of Loreto.

MISIÓN SAN FRANCISCO JAVIER DE VIGGE

116.6   Loreto International Airport. The highway continues south along the narrow coastal plain between the abrupt escarpment of the Sierra la Giganta and the warm waters of the Gulf. The escarpment was a formidable barrier to the early missionaries. Isla del Carmen dominates the Gulf and the northern portion of uninhabited Isla Danzante is just visible to the southeast.

The highway was constructed in the Gulf Fault Zone. The disjointed and irregular hills in the foreground are in the fault zone. The crushed rocks adjacent to the highway lack the impressive flat bedding that is well exposed in the high eastern escarpment beyond the fault zone to the west.

110.9   This turnoff leads to Nopoló Beach Resort with a golf course. A beautiful cove and mangrove swamp are located near a tómbolo that connects the peninsula to a small rocky island. At some high tides this tómbolo may be awash. The scenic beach in this cove is a rare lithic sand. The highway ascends a small grade through basalts with a lagoon, inlet, and the small mangrove swamp of Nopoló. A number of inhabited caves have been found in the Sierra la Giganta to the east of Nopoló. They were occupied by Indians who subsisted on the clams and marine life of the cove.

PANGA FISHING AT NOPOLÓ - CAN YOU SEE THE ANIMAL HEAD?

At Nopoló, winter visitors may be treated to a spectacle of wildly-diving Brown Pelicans. During the winter, currents and winds combine to pile up schools of small fish along the coast of this area. The pelicans take advantage of the conditions and become embroiled in feeding frenzies.

INDIAN CAVE AT NOPOLÓ

The **California Brown Pelican** is a large aquatic fish-eating bird along Baja's coasts and on many of the Gulf islands in Baja where they nest in large colonies. They are seen flying in straight lines only inches above the surface of the water. Feeding is accomplished by diving into the sea from as high as 150 feet. The slow deliberate flight of the brown pelican, over the water, with sudden plunges for fish, makes its identity unmistakable. The Brown Pelican is commonly seen on piers and docks.

107.4   Nopoló Golf course.

103   To the left is Playa Notri, another lithic sand beach where many species of game fish, which include Marlin, Sailfish, Sierra, Dolphin Fish, Yellowtail, and Grouper, often come close to shore.

Many species of birds inhabit the Gulf Coast Desert. Some of the more common birds of this area are listed below:

BIRD NAME	LIKELY LOCATION
Amer. White Pelican	Gliding along the shore
American Kestrel	Wires and fence posts
Cactus Wren	On cacti
Calif. Brown Pelican	Gliding along the shore
California Quail	On the ground
Costa's Hummingbird	Feeding on red flowers
Gila Woodpecker	In the scrub
Gray Thrasher	On ground looking for food in litter
Greater Roadrunner	Crossing the highway
Ladder-back Woodpecker	Flitting in the air
Pyrrhuloxia	Mesquite and oak woodlands
Red-Tailed Hawk	Tops of poles and fence posts
Turkey Vultures	Soaring in the sky
Western Meadowlark	Fence posts and fence wires
Xantu's Hummingbird	Feeding on tubular flowers

99.5   Isla Danzante in the distance to the south is composed of faulted Miocene andesite breccias and volcaniclastic sedimentary rocks. The north end of Isla Monserrat, composed of Miocene volcanic rocks and Pliocene marine sedimentary rocks, and Isla Santa Catalina to the left, composed of Mesozoic granitic rocks are in the distance. The largest island, Isla del Carmen, composed of the Miocene volcaniclastic and Pliocene marine rocks, is directly offshore. To the north is the basalt cone of Isla Coronado.

99.3   There is a large solitary Fig Tree (*Ficus palmeri*) growing on the beach at the end of the gravel bar to the left.

WILD FIGS are members of the Mulberry family. They range from the palm oasis of San Ignacio south to the Cape region where they are usually seen growing alone on rocky cliffs, in canyons, and occasionally on gravel beaches such as this one. Wild Figs produce a barely edible, dry fig in the late spring and early summer.

97.5   This is the entrance to Juncalito Cove. This is a great place to stop and watch for birds. One that is especially nice is the Crimson Pyrrhuloxia.

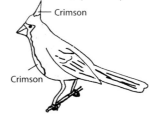

The **Crimson Pyrrhuloxia** inhabits the southern half of Baja where they are known as the gray bird of the cardon cactus. They are a beautiful crimson and gray colored cardinal relative with a yellow, straight parrot-like bill. Pyrrhuloxias are commonly seen in the mesquite scrub along the road to the Gulf's Bahía San Luis Gonzaga and surrounding Loreto where they usually feed on the ground, taking seeds and insects. This bird derives its name from the Greek, meaning crooked-billed red finch.

The **Magnificent Frigate Birds** are the most aerial of all sea birds and thus present a common but thrilling sight over the waters along the gulf coast of Baja. These birds possess the greatest wingspan (up to 8 feet) in proportion to their body of any known species and are hardly equaled by any bird in sustained, soaring flight. Their identifiable field characteristics are the prominent crook in the slender streamlined wing, the long slender forked tail, the way they soar high in the sky seemingly motionless, their steep, swift dives to snatch fish from the sea, and the naked red inflatable throat

(gular sac) of breeding males. Immature birds have white on their heads and throats. They often act in an aggressive and piratical manner, chasing and colliding with terns and gulls, upsetting these birds so that they regurgitate their last fishy meal which the Frigate will then aerobatically snatch from midair. The name Frigate (a pirate ship) was derived from this behavior. Other methods of obtaining food are to steal chicks from colonies or, as a last resort, to fish for themselves. Frigates nest on islands in large, untidy stick structures in anything from bush to a mangrove tree to avoid predation by four-footed predators.

94.3   **PUERTO ESCONDIDO:** This side road leads 2.4 kilometers to Puerto Escondido. Steinbeck in his *Log of the Sea of Cortez* claimed this as his favorite place in the Gulf. The authors of this guide have spent weeks here and agree. This area is worth a visit. The coastal waters abound with fish and other sea life. Twenty-five pound Yellowtail have been caught off the pier. A fault Perpendicular to your view is exposed in the saddle directly in front of the pier with dark basalts on the near side and fluviatile andesite breccias on the far side.

93   The offshore view is dominated by Isla Danzante. Isla Carmen is to the left and Monserrat is to the right. The three small isolated "rocks" in the bay to the right are Los Candaleros (the Candles).

SCARP ALONG GULF FAULT ZONE

89   The view to the rear is of the uplifted steep eastern escarpment of the Sierra la Giganta. The flat and well-bedded strata extend from the Pacific eastward across the peninsula to the edge of the scarp. In the fault zone, the beds become chaotic and crushed.

86   The highway continues south along cliffs of fluvial volcanic breccia exhibiting large clasts and numerous caves.

The flora of this area is dominated by *Acacia*, Garambullo, Pitaya Agria, Pitaya Dulce, Cardon, Mistletoe, Palo Blanco, Palo Adan, and the Vining Bougainvillea-like San Miguel. White-winged Doves (*Zenaida asiatica*) are commonly seen on the ground or flying across the highway.

White-Winged Doves are the only doves with large white wing patches that contrast with the olive-brown colors of their body. Although they are widespread throughout the peninsula, they are most commonly seen in thorn scrub, mesquite groves, Riparian Woodlands, and the open dry desert areas south of Loreto. They have a drawn out, cooing call "who-cooks-for-you" in many variations.

84.2   Turnoff to Ligui and a beautiful cove with a rare lithic sand bottom.

83.5   The highway turns through a wash, reaches the bottom of the grade, and begins the ascent of the steep grade south of Ligui. Many fault shears of the Gulf fault zone are exposed in the road cuts on this grade.

Between here and the top of the steep grade, the highway passes over a segment of the major Gulf Fault Zone. Most of the major lateral motion along this fault was pre-Miocene. There is evidence of lesser vertical Miocene to Recent motion. The zone is difficult to identify because it is wide and diffuse. The complexly faulted areas, offset, and deformed bedding of this region is generally marked by a foothill belt rather than the undeformed well-defined bedding of the mesas of the Pacific slope.

76   This turnout provides a view to the north of the majestic of the Sierra la Giganta. The high crest of the range with the undeformed sedimentary rocks that extend from the Pacific slopes drops rapidly to the east into the undulating fault-deformed foothills adjacent to the highway.

70.8   As the highway crests the Ligui grade, a road to the east leads to Microondas Ligui. The highway descends, for tens of kilometers, through an east-west canyon down the gentle western slope of the Sierra la Giganta.

GULF FAULT ZONE AT TOP OF GRADE

69   The undeformed well-bedded strata of the Pacific drainage of the Sierras are now very obvious along the canyon walls.

UNDEFORMED WELL-BEDDED STRATA OF THE PACIFIC DRAINAGE

64   This side road leads 41 Km to the gulf at Bahía Agua Verde and Punta San Marcial. The cliffs on the left are covered with two kinds of lichen (Fungus with an intracellular mutualistic single celled algae). The most abundant is green; the less obvious are orange-red.

60   The Cardon cardonal is extensive and primarily located on the

warmer-dryer south facing slopes, while Mesquite grows abundantly in the well-drained canyon bottom. The two sides of the canyon show notable differences in vegetation typical of north and south-facing slopes.

56   The vegetation of this region represents an ecotonal zone *(See 2:141)* of a mixture of plants of the Magdalena Plain desert area and the Gulf Coast desert area. It is dominated by a Cardon-Palo Verde forest with scattered Pitaya Dulce, Acacia, Cholla, Beaver-tail, vining San Miguel, Palo Blanco, Palo Adan, Purplebush, Lomboy, and Elephant Tree.

54.5   On the hill to the south, brown basalt caps the lighter gray-brown andesite breccias, lahars, and fluvial sedimentary rocks of the Comondú Formation. They dip gently to the west. The highway descends at approximately the same gradient as the arroyo. The gradient is controlled largely by the slight dip of the Comondú Formation toward the Pacific Ocean. The arroyo descends for some distance along a single bed, drops down through a few beds, and then flows along another bed. Marine Miocene rocks are not exposed along the highway as it crosses the peninsula in this region.

49   Andesite breccias are exposed in the wall of the canyon. These monolithologic andesites take the form of lahars with flows, flow conglomerates, and fluvatile conglomerates grading into sandstones and well-bedded sedimentary rocks. This is a facies of the Comondú Formation.

45   The highway leaves the canyon and passes onto the Magdalena Plain. It is largely composed of Miocene and Pliocene rocks overlain by the same type of Pliocene-Pleistocene limestone that caps the Vizcaino Plain to the northwest. These sedimentary rocks are only occasionally exposed in the arroyos or road material quarries of the Magdalena Plain. Bahía Magdalena is a deeper remnant of this shallow embayment.

On a clear day, the Pacific islands of Isla Margarita and Isla Magdalena, composed of Franciscan metamorphic rocks, are visible. These islands are remnants of the Pacific Plate sea floor that was subducted under the North American Plate. Radiometric ages from these rocks cluster around 135 m.y. They are similar in lithology to the Coast Ranges of California.

43   The vegetation of this region is characteristic of the dry Magdalena Plain region of the Desert Phylogeographic Province. This extremely dry region of Baja receives 2 to 5 cms. of precipitation annually.

Magdalena Plains Subregion. The Magdalena Plains have an abundance of cacti such as Cardon, Cholla, Garambullo, Candelabra Cactus, and Creeping Devil Cactus. Palo Adan, Palo Blanco, Palo Verde, and San Miguel vine are found in the foothills. Mesquite groves grow in the arroyos.

167

18 The "fog desert" of this area encourages lichens and Gallitos on the shrubs and cacti. Rocella *(Ramalina* species) is the most noticeable lichen. Other common plants are Datilillo, Lomboy, Maguey, Pitaya Agria, Pitaya Dulce, and Torote.

28   Zona de Neblina or "Zone of the fog." The fog moisture in this zone enables an abundance of epiphytic gray-green foliose lichens and dark Ball Moss to grow on the shrubs and cacti.

20   This area is much drier than that near Kilometer 43 and supports significantly fewer, shorter scattered Cardons, Pitaya Dulce, Candelabra Cactus, Ball Moss, Creosote Bush, Lomboy, Palo Verde, and Palo Adan.

16.5   The size of the Que Reparrio bridge attests to the fact that on some occasions, a large quantity of water flows over this plain.

0   As the highway intersects the Highway to Ciudad Insurgentes. Turn sharply to the southeast. The kilometer markings change to 236 at this junction and descend as the highway approaches La Paz.

This area is an intensively farmed salad bowl. However, there is a limited supply of water for irrigation. Thus, there are many fallow fields interspersed between the natural vegetation and the active areas being farmed.

223   There are three large bridges over a wash that also attest to the fact that on some occasions, a large quantity of water flows over this plain.

212   **Ciudad Constitución** is a thriving agricultural community and is the gateway to Bahía Magdalena and Puerto San Carlos.

Log 8 - Constitución to La Paz [212 kms = 132 miles]

The highway crosses the flat Magdalena Plain, with views of the offshore islands, to Santa Rita. From this point, the highway begins to traverse a series of gentle washes and dissected mesas in marine sedimentary rocks to San Agustin where it climbs onto a dissected surface of fluvial sedimentary rocks and crosses the peninsula. Near La Paz, the highway descends the Gulf Scarp through fluvial volcanics to the alluviated La Paz Plain.

212   **CONSTITUCIÓN** is a large agricultural community.

Farming has been made possible by **deep artesian wells** drilled by the Mexican Government that bring "fossil water" to the surface from ancient aquifers. The amount of water withdrawn from these aquifers is carefully monitored to

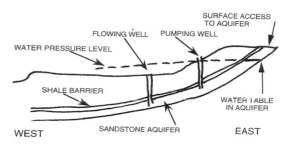

prevent the over usage and eventual depletion of this limited resource. Some water is naturally available from infrequent summer rainstorms or late summer-early fall chabascos or hurricanes. These produce sudden downpours that may overflow the arroyos in a matter of minutes. However, much of this water is not available to the flora of the Magdalena since it runs off quickly before much of it has a chance to soak to the roots of the plants.

198   The village of Via Morelos is located in the middle of a farming area. Most of the nonagricultural regions of the Magdalena Plain support disturbed vegetation because the land has either been cleared or overgrazed.

194.7   Misión San Luis Gonzaga is located in the low foothills east of the Magdalena Plain. The mission was formerly the site of a garrison of troops used to reinforce the sailors on the Manila galleon when it came through the Bahía Magdalena area. This mission is situated along a stream that follows a transpeninsular fault offsetting the Cretaceous geosyncline by a total of 20 miles in a right lateral sense. The fault has deformed lower Miocene sedimentary rocks and is overlain by flat-lying 20-million-year-old volcanic tuffs. A high clearance vehicle can reach the mission from this junction.

173.8   The quarry on the right exposes limestones of the fossiliferous Miocene-Pliocene Salada Formation. There is a graded road just south of the quarry that leads 32 km to Cancun on Bahía Magdalena.

MISIÓN SAN LUIS GONZAGA

MILITARY BARRACKS AT MISIÓN SAN LUIS GONZAGA

This plain resembles the Great Valley of California. Structurally, the two regions are identical. The highway roughly follows the axis of the Cretaceous geosyncline, as defined by geophysical surveys, and is underlain by up to 60,000 feet of Cretaceous and Tertiary sedimentary rocks (*See* 5:162).

168   The vegetational cover between here and Santa Rita is sparse. Depending on the location, the flora is alternately dominated by Cardon, Cholla, Elephant Tree, Palo Verde, Palo Adan, Cheesebush, and Creeping Devil Cactus, Leatherplant, Greasewood, Garambullo, and Pitaya Dulce.

On clear days, the islands of Isla Santa Margarita and Isla Magdalena are seen to the west across Bahía Magdalena. These islands are composed of Franciscan-type rocks similar to the Coast Ranges of California. They contain pillow basalts, serpentinite, and garnet hornblendites. These rocks are K/Ar dated at 134 m.y. (Forman, 1971; Yeats, *et al.*, 1971).

157   From Santa Rita, the highway traverses through a series of washes and dissected mesas of the gentle Pacific slope drainages. The washes are populated by dense mesquite, while the slopes support Creosote, Leatherplant, and Palo Adan with stands of Cholla and Pitaya Dulce.

145   This road cut exposes limestone that has been mapped as part of the Miocene-Pliocene Salada Formation. The Salada Formation (Heim, 1922) is named for marine Pliocene and Pleistocene sedimentary rocks in Arroyo Salada. Judy Smith (Smithsonian) now feels that the type section of the Salada Formation ranges from late Miocene to early Pliocene in age and represents a shallow marine embayment.

EL RIFLE

135.2   The highway makes a long crossing of an arroyo. Shark teeth have been found in the Miocene exposures located near the El Rifle microwave tower to the left of the highway.

The floral species that cover the Pacific slopes of the Sierra de la Giganta are typical desert dominants of the Magdalena Plain area of the Desert Phytogeographic Region. They continue to be represented by epiphytic Ball Moss and *Ramalina* lichen that grow on the elephant trees *Bursera microphylla* and *B. hindsiana*, Palo Adan, Leatherplant, Cardon, Jumping Cholla, Pitaya Dulce, and Garambullo.

127.1   Turnoff to Rancho San Pedro de la Presa. The road heads east to Misión La Presa, north to Misión San Luis Gonzaga, and back to Ciudad Constitución. It traverses a section of Miocene rocks in this part of the peninsula.

126   The bridge at Puente Ventura was washed out in a flash flood in 1978. Future washouts may occur since the highway crosses a wash that periodically and briefly conducts enormous amounts of water from the infrequent sudden downpours and chabascos that are common in this region

112   The prominent peak to the east is Cerro Colorado.

CERRO COLORADO

Mesas capped by dark brown and gray-green volcanic rocks can be seen to the left. Immediately below, sandy yellow sedimentary rocks are visible in the middle of the section, and shaley yellow beds are at the base of the section. These rocks were originally referred to as the Isidro, San Gregorio and Monterey Formations by earlier workers. They are now renamed and are regarded as members of the El Cien Formation of Applegate.

These Miocene formations are well exposed in various cuts in arroyos over large parts of the Magdalena Plain. The plain is so flat that few rocks are exposed outside of the arroyos.

108   For the next several kilometers, the highway passes through sandy yellow beds that have been mapped as the Isidro Formation. These are near-shore facies, but not quite as near-shore nor as tuffaceous as the San

172

Ignacio Formation seen to the north near San Ignacio. Deeper water shaley beds locally interfinger with the sandy yellow beds. They are often not actually mappable as separate units.

100   The highway climbs through the beds of the El Cien Formation and passes through the small settlement of El Cien. On July 11, 1991, The authors watched a spectacular total eclipse of the sun from this location.

On the left side is the 1,000 foot section at Cerro Colorado that is at the core of a major anticline. Most of the exposures on Cerro Colorado are the Oligo-Miocene marine El Cien Formation with a nonmarine tuffaceous redbed near the top. The tuffaceous redbeds and a conglomerate bed on Cerro Colorado can be traced across the peninsula into the "Comondú" Formation on the Gulf. Redbeds and fluvial sedimentary rocks become thicker to the east; as the marine Miocene rocks at the base of the hills become thinner indicating an eastern source for the volcanics. In the distance, the "Comondú" Formation can be seen on the south limb of the anticline where progressively younger beds of the Comondú Formation are exposed.

The name **Comondú Formation** (Heim, 1922) has been applied to a diverse assemblage of volcanic and volcaniclastic rocks of Miocene to recent age in Baja California.

The **Isidro Formation** was defined by Heim (1922) as exposures of greenish, whitish, and yellowish sandstones with interbedded greenish shales in the La Purisima area. The Isidro Formation is Early Miocene and overlain by a 22-million-year-old tuff in the Cerro Colorado area.

The widespread **Monterey Formation** (now renamed the El Cien Formation) and San Gregorio Formations were also defined by Heim (1922) for exposures of hard, clear, siliceous, diatomaceous yellowish shales in the La Purisima area and in the Magdalena Plain.

97   The highway passes through exposures of the El Cien Formation for the next three kilometers. Abundant splintered pieces of siliceous shale scattered over the surface indicate the presence of this formation.

92   The Magdalena Plain desert vegetation continues to be dominated by species of Jumping Cholla, Teddy Bear Cholla, Cardon, Elephant Trees, Creosote Bush, Pitaya Dulce, and Pitaya Agria. The Wash Woodland vegetation continues to be predominated by scattered specimens of the Acacia trees and Palo Verde. Another interesting and vicious looking plant of this region is the Creeping Devil *(Machaeroocereus eruca).*

ACACIA: There are more than 900 species of Acacia trees and shrubs in the world with 64 species found in Mexico. About 20 species of Acacia occur in Baja from the northeastern Desert Region to the Cape Region. The fruit of the Acacia is a pod (specialized folded leaf) with several seeds. Indians ground the seeds into a meal. Cattle also relish the pods.

CREEPING DEVIL is a prostrately growing cactus because of the weight of the heavy stems. This sharp spined, tire menacing cactus sends out a dense network of branches that often produce adventitious roots where they touch the ground. As the prostrate branches slowly grow forward, the older hind parts die; this plant spreads and multiplies itself by asexual, vegetative reproduction. This mechanism of reproduction is often used by species that inhabit arid environments as a way to avoid the "water expensive" process of sexual reproduction that involves the production of moist flowers and seeds that require water for germination. Creeping Devil grows in alluvial soils over the Llanos de Magdalena.

89   The cliffs to the north offer a cross-section of the geology of this part of the peninsula. The upper thick pink tuffs and gray-green fluvial sandstones are the Comondú Formation; the middle yellow-gray sandy San Ignacio Formation is visible to the east. It grades downward into the yellow, marine sedimentary rocks of the Monterey and Isidro Formations on the west side. The chevron folds in this road cut are in the Monterey Formation.

FOLD IN MONTEREY FORMATION

85.5 The contact between the brownish marine Paleocene Tepetate Formation and the overlying yellowish marine Oligocene-Miocene El Cien Formation is exposed on the hill to the north of the highway. The flat-lying contact between the formations is a slight angular unconformity. The rocks of the Eocene part of the Tepetate Formation were tilted and partially removed during a period of uplift and erosion producing the unconformity.

84 Exposures of redbeds and yellow-brown sandstones of the Tepetate Formation. The highway road cut has exposed a discocyclinid Coquina. Discocyclinids are button like foraminifera that grow to one centimeter in size.

83 The Pacific Ocean can be seen to the right. Some of the distant small mesas are composed of Pliocene Salada Formation that overlies the yellow-brown Tepetate Formation.

79 The hills to the left contain the Tepetate Formation. The highway descends into an amphitheater with exposures of the flat well-bedded strata of the Tepetate Formation. To the north are the brown Paleocene marine Tepetate Formation and the overlying yellowish El Cien Formation.

78.5 Along the highway, the surface of the hills are littered with small flat shale fragments called "tepetates" (slabs).

77 Arroyo Conejo. Exposures of the Tepetate Formation along sides the highway. Walk along the east bank of the arroyo to view this formation.

Knappe (1974) stated that in Arroyo Conejo "the **Tepetate Formation** consists of a series of sandstones and shales. Foraminifera collected from a section of this formation in Arroyo Conejo are Eocene in age and correlate with assemblages found in California, Oregon, Washington, and elsewhere. The oldest and youngest benthonic fauna examined correspond to the Penutian and Ulatisian stages of Mallory (1959).

Carreno and others (2014) indicated that "based on the planktonic Foraminifera, *Acarinina pentacamerata* (latest lower Eocene) and *Hantkenina nuttalli* (earliest middle Eocene) zones were recognized that indicate an age range from 51.2 to 48.4 Ma. Benthic Foraminifera suggest correlation with the Penutian to Narizian California stages, which agrees with the planktonic foraminiferal biozones. The sedimentary structures and microfossils indicate deposition within a transgressive- regressive- transgressive cycle, including environments from the inner to outer marine shelf in depths shallower than 150 m. Gravity flows dominate the sequence, and storm-deposited beds are present in the middle of the section."

76   Rancho San Agustin is located on a river terrace next to the arroyo.

74.2   The highway begins the ascent of a grade through the sandstones and shales of the Tepetate Formation.

74.2   To the right of the highway, a road material quarry exposes Tepetate sandstones and shales with discocyclinid forams.

70   The El Cien Formation is exposed in this road cut. The strata along the highway are near the contact between the tuffaceous fluvial Miocene Comondú sedimentary rocks and marine rocks. In the canyon to the left are exposures of the light-yellow to gray marine Miocene rocks. In the near distance are the overlying red tuffs and gray fluvial sedimentary rocks of the Comondú Formation. The El Cien Formation is represented by well-bedded siliceous and tuffaceous sedimentary rocks that include shales, siltstones, and cherts. The marine Miocene rocks on the Pacific side of the peninsula are stratigraphically   correlatable with nonmarine fluviatile and tuffaceous beds of the "Comondú" along the Gulf of California.

67   To the southwest is the Pacific Ocean and the low mesas underlain by the Salada Fm. The canyons along the highway contain abundant exposures of light greenish to yellowish-gray rocks of the El Cien Formation.

67.5   Rancho Aguajito.

The highway travels along the tilted erosion surface of the gentle Pacific slope and passes through the Miocene fluviatile volcaniclastic sandstones of the Comondú. The Highway follows this surface to Km 34 on the Gulf

176

escarpment. In this 30 Kilometer stretch, all of the road cuts are in the fluviatile sedimentary rocks of the Comondú Formation which John nicknamed the "gray-green grunge."

62   The dense, dark evergreen trees with gray trunks scattered in this area are Palo San Juan (*Forchammeria watsonii*). It is a member of the Caper family (Capparidaeae) that produces an edible deep-purple fruit.

The vegetation consists of Lomboy, Elephant Trees, Datilillo, Greasewood, Pitaya Agria, Pitaya Dulce, Cardon, Hedgehog Cactus, Palo Adan, Palo Blanco, *Acacias*, Jumping Cholla, Teddy Bear Cholla, and Mistletoe.

57.2   Outcrops of Fluvial grey volcaniclastic sandstones are exposed in this area.

Along this stretch of the highway, Turkey Vultures (*Cathartes aura*) are seen roosting on the Cardon Cactus and in other trees. Or they are seen soaring in wide circles, their black wings in a broad "V" rocking quickly from side to side.

Turkey   Vultures (buzzards) are ubiquitous throughout Baja and are protected by law for their value as "cleaner-uppers" of dead   bodies.   Turkey Vultures are naked-headed, carrion   eaters   that   are commonly seen scavenging on the ground and along roadsides. Along highways, one may see a group of them devouring a dead animal that was detected by sight or smell while soaring high in Baja's skies. In flight, they appear to rock from side to side and seldom flap their wings. They are often seen perching on the tips of Cardon branches "sunning" themselves.

36.1   Microondas Matape.

34 On a clear day, the view from the top of the grade is impressive. The viewpoint is on the south limb of a major anticline in the Miocene rocks that dips south under Llanos de Todos Santos. To the northeast are

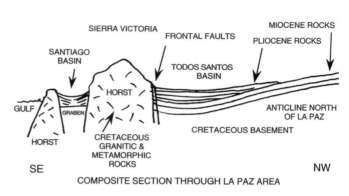

SIERRA VICTORIA
FRONTAL FAULTS
MIOCENE ROCKS
PLIOCENE ROCKS
SANTIAGO BASIN
TODOS SANTOS BASIN
HORST
GULF
GRABEN
ANTICLINE NORTH OF LA PAZ
CRETACEOUS BASEMENT
HORST
CRETACEOUS GRANITIC & METAMORPHIC ROCKS
SE
NW
COMPOSITE SECTION THROUGH LA PAZ AREA

Islas Partida and Espirito Santo; they are composed of highly faulted and tilted volcanic rocks of the Comondú Formation. To the east, the rugged granitic spine of the Sierra de la Victoria (65-75 million years) begins on the peninsula east of La Paz and stretches south to the Cape Region. This range is separated from Llanos de Todos Santos by a major fault zone.

The avifauna of this region is not diverse. Species commonly seen are:	
American Kestrel	Wires and fence posts
American Robin	Woodland openings, forest borders
Belding's Yellowthroat	Stays low in grassy fields, marshes
Blue-Gray Gnat Catcher	Perched on vegetation
California Quail	On the ground
Common Ground Dove	On the ground
Costa's Hummingbird	On red or yellow flowers
Gila Woodpecker	In the scrub
Gray Thrasher	Looking in ground litter for food
Ladder-Back Woodpecker	Flitting in the air
Loggerhead Shrike	Perched on wires, fences
Northern Cardinal	Woodlands, streamside
Pyrrhuloxia	Common in thorny bushes, mesquite
Turkey Vultures	In sky looking for carrion
Varied Bunting	Perching on vegetation
Western Meadowlark	Fence posts and wires
White-winged Dove	Dense mesquite, riparian woodlands, desert
Xantu's Hummingbird	Feeding at tubular flowers
Yellow-billed Cuckoo	Flying through vegetation
Yellow-eyed Junco	Coniferous & oak slopes
Yellow Warbler	Wet areas, woodlands

Mexican Chickadees, White-winged Doves, Cactus Wrens, Gila Woodpeckers and Falcons are commonly seen in this area.

The **Peregrine Falcon** is distinguished by a "helmet" formed from a black head, neck, and wedge extending below the eye. This large-sized bird flies fast and rarely soars. Peregrine falcons inhabit open wetlands near cliffs and are regularly seen at coastal bays and lagoons in winter. They mainly prey on ducks, shorebirds, and seabirds.

28   This area is an ecotone between the flora of the Magdalena Plain subdivision of the Desert Phytogeographic Region and the Cape Phytogeographic Region. Due to overgrazing, Cholla has become a prominent element of the understory vegetation. Specimens of Agave, Pitaya Agria, Pitaya Dulce, Cardon, Acacia, Elephant Tree, Lomboy, Palo Blanco, and Palo Adan are growing on the hillsides. The Creosote Bush and Mesquite are the dominant plants of the Wash Woodland flora.

25   The view to the left rear from a low rise is of the pink tuffs and gray-green fluviatile sedimentary rocks in the south limb of the San Juan anticline that rises to the north and dips out of sight to the south. The highway crosses the anticline in fluvial volcaniclastic sandstones above the highest pink tuff. These rocks were deposited prior to the opening of the Gulf.

The vegetation consists of Lomboy, Palo Blanco, Palo Adan, Mesquite, Elephant Tree, Cardon, Pitaya Agria, Pitaya Dulce, Cholla, and Agave. This area is an ecotone between the flora of the Magdalena Plain subdivision of the Desert Phytogeographic Region and the Cape Phytogeographic Region.

COMONDÚ FORMATION ON GULF

17   The side road leads to the phosphate mining of San Juan de la Costa. Phosphates are mined and shipped in Mexico for use as a fertilizer. This mine supplies about 40% of Mexico's phosphate needs. About 90% of the world's production of phosphorous is from sedimentary phosphate rock (phosphorite) of marine origin such as that mined in this area. Late Oligocene fossils including Mexico's oldest whales, ancestors of baleen whales that still had teeth are found in the San Juan member of the El Cien Formation.

15   The highway passes through the bayshore area of El Centenario.

Bahía De La Paz, the largest bay on the peninsula's east coast, and the muddy sand spit of El Magote are visible. El Magote is formed by the southward transport of sand by currents along the eastern coast of the peninsula. It separates the bay from Ensenada de Los Aripes, the shallow bay to the west.

8.9   Turnoff to the right to the La Paz International Airport.

6.8   The right fork at the overpass is the most direct route to the cape. Proceed just past shopping Center to Avenida Santa Ana, then right on Forjadores which is the main highway to the cape. There are at least two signed streets in town which direct traffic toward Cabo San Lucas. To reach Pichilingue continue to the left on Abasolo to the Malacon.

0   **LA PAZ:** The Malacon at La Paz is a coconut palm-lined sea wall and pedestrian walkway combination that borders the bay in the main part of La Paz. At the intersection of 5 de Febrero, Highway 1 becomes Mexico 11 and bends to the north along the bay toward Pichilingue. It is well worth the side trip to Pichilingue to see the beautiful coves and biological sand beaches (See the cover of the book).

SIDE TRIP TO PICHILINGUE AND THE FERRY DOCK VIA MEXICO 11:

Mexico 11 to Pichilingue wraps around Ensenada de Los Aripes and leaves by way of Paseo Alvaro Obregon. There are a number of groins built along the beach to slow the drifting of sand due to currents that flow into the bay. At the time of this log, this area was still undergoing development with new and unpaved roads. Some roads are now divided.

1.5   The road travels through the pink tuffs and the gray-green volcaniclastic sedimentary rocks, massive fluviatile bouldery conglomerates, and mega-breccias of the Comondú Formation.

2   Watch for herons, egrets, storks and other marsh birds.

The **Cape Phytogeographic Region** is vegetated by a dense thorny, Acacia dominated flora known as Thorn Scrub. Thorn Scrub grows where summer precipitation exceeds 300 mm/Yr.

2.8   The highway climbs a small grade through faulted pink tuffs and fluvial volcaniclastic sandstones of the Miocene Comondú Formation. The road crosses several steeply dipping north-south normal faults related to the frontal faulting of the Sierra Victoria.

4.1   The highway passes the coralline "sand" swimming beaches of Coromuel. A native wild fig tree (*Ficus palmeri*) is across from Coromuel beach. Rainfall, scarce in La Paz, arrives in the form of violent tropical storms called Chabascos. As a result, summers are hot and dry. The name "Coromuel" refers to the cooling wind from the cooler Pacific Ocean to the south across the peninsula and refreshes every summer afternoon.

4.5   The ship channel markings delineate a narrow channel very close to the near shoreline. The southward movement of sand along the Gulf has formed El Magote and piled sand in the mouth of the Ensenada. Tidal currents scour the channel as water moves in and out of the Ensenada.

5   The Las Conchas Resort is located at El Caimancito. The small offshore rock, Islita Caimancito resembles a lurking alligator (Caiman).

The highway passes through volcaniclastic rocks and tuffs of the Comondú Formation which is exposed around the resort. On a clear day, the major San Juan anticline in the Comondú Formation can be seen on the west side of Bahía de la Paz. This anticline and one to the north stretch for over 200 kilometers north of La Paz. The same bed can be traced for nearly the entire distance. These anticlines are strong petroleum prospects as the Cretaceous geosyncline underlies the volcanics to the eastern edge of the peninsula. There are exposures of pink tuffs for the next several kilometers.

Because of unfavorable edaphic (soil) features, the vegetation is very sparse consisting of Cardon, Lomboy, Palo Adan, *Acacia*, and Candelabra.

6   Road forks. The left fork follows the old route to Punta Prieta. Costa Baja marina and resort have been built here. This log follows the left fork.

9   Punta Prieta (dark point) is the location of the government owned and operated Pemex oil docks and storage tanks. Oil is shipped from mainland Mexico for storage and distribution. Good exposure of the

conglomerates of the Comondú Formation. The road bends around the point and joins with the new road that was the left fork at Kilometer 6

9.4   View of the first of a series of shallow coralline biological "sand" coves with azure blue waters backed by a Mangrove swamp and a small estuary. The beach is formed by light-colored coralline biological "sand."

11.8   The highway skirts another cove with exposures of very large mega-breccias of the Comondú Formation, then cuts through a mega-breccia.

The water in this beautiful cove is very blue-green in stark contrast to the light-colored coralline biological "sand" beach. Great Blue Herons are often seen in the Mangrove swamp located along the shore of the cove.

The **Great Blue Heron** is often seen in the marshes. It stands motionless or advances slowly one step at a time, lifting each foot stealthily from the shallows without a ripple. Herons may stand still as a statue for over half an hour while waiting for prey. With a lightning-quick forward lunge of their long neck and bill, it will capture its prey. It prefers fish but will also eat birds, insects, snakes, frogs, and crustaceans. A Great Blue Heron is  recognized by its long snakelike neck held in an "S" shaped curve and its slow wing strokes, long legs, and nearly 6-foot wing span.

14.5   The highway skirts a cove through a Mangrove marsh and traverses a super-tidal flat. The vegetation on the supper-tidal flats is sparse and impoverished due to the highly saline soil (*See* 16:35). Cardons are seen on the hillsides along with scattered Lomboy, Organ-pipe and Mission Cacti, and Acacia. A small cardonal covers the high flat behind the lagoon.

15.4   The highway passes through a nearly vertical road cut of massive resistant fluvial mega-breccias. Isla Pichilingue is across the lagoon.

16   Traffic circle. This is the Ciencias Del Mar, Departmento de Geologia Marina Labs and Research facilities branch of the Universidad Autonoma de Baja California, Sur (UABCS) on the tidal flats.

16.5   The ferry dock with several loading areas is to the left. The rocks in the road cut are andesite mega-breccias.

17.2   Bahía Pichilingue is a deep-water port for La Paz. It avoids the narrow and difficult passage into the main harbor. Continue straight ahead to Balandras and Tecolote.

BAJA'S FIRST SETTLEMENT: Isla Pichilingue is the site of the first attempt in 1535 to settle Baja California by Herman Cortez, the conqueror of Mexico. However, supply problems resulted in the failure of that first settlement.

18   The road passes a coralline sand beach, climbs through andesite breccias, then through a dense cardonal with Elephant Trees and Mesquite.

23.8   Junction. The right fork goes 2 Kms. to the beaches of Tecalote. The left fork goes 1 kilometer to Balandras Bay, one of the most beautiful bays in Baja. The cover photo was taken from the ridge on the south side of the bay.

Balandras Bay with *Isla Espíritu Santos on horizon*

Balandras Bay – *How peaceful it is. The balandra is on the far point.*

You can wade to the right around the small rocky point (it's shallow), walk the beach to the next small point, and then wade to the end of that point to see the Sombrilla [umbrella].

SOMBRILLA AT BALANDRAS

SOLUTION-CUT NOTCH is produced where weathering of the hillsides is very slow. The rocks on the shoreline are moistened by salt spray. As the water evaporates, the salt remains and crystallizes in the pore spaces and cracks, prying the rock apart to leave a notch just above the water line

SOLUTION CUT NOTCH

Where To Now? Return to La Paz and continue to the south on Highway 1 using the La Paz to Cabo San Lucas log 9 for your guide.

Log 9 - La Paz to Cabo San Lucas [211 kms = 131 miles]

The highway follows the Gulf fault zone in an alluviated syncline. It then turns to climb over uplifted granitic and metamorphic horsts (El Triunfo, then San Antonio and San Bartolo) to descend to the Gulf at Buena Vista. The road then travels south on marine and nonmarine sedimentary rocks, along the relatively flat Santiago Trough between two high granitic horsts to San José de Cabo. From here, the highway skirts the southern edge of the Sea of Cortez on alluvial fans to Cabo San Lucas and its granitic headland.

From the malacon, the best route out of La Paz toward the Cape is south on 5 de Febrero, which is a divided street, then diagonally right on Forjadores.

210.5   Universidad Autonoma de Baja California Sur is on the right.

South of La Paz, the edge of the Sierra la Victoria range is remarkably straight as a result of the recency of faulting in this area. The highway follows the Todos Santos graben-syncline that is situated between the Sierra la Victoria Fault Zone and the south limb of the anticline in the Tertiary sedimentary rocks. This graben is filled with Pliocene to Recent sediments.

The Mesozoic granitic and metamorphic rocks of the Sierra la Victoria are similar in age, relationships, and rock type to the Peninsular Range Batholith. They were part of the same granitic belt prior to the Cenozoic extension that stretched Baja and separated it from the mainland. Age relationships in the Gulf indicate that the stretching took place in the Middle Miocene and the separation took place at the mouth of the Gulf of California about 4-5 million years ago and continues today.

186   The highway enters the **Cape Phytogeographic Region of Baja**. The predominant plants of the Cape Region flora in this area consist of Cardon cactus and a mixed "forest" of leguminous trees such as Palo Mauto, Palo Blanco, Acacia, Screwbean Mesquite, and Palo Verde. The dominant plant along the highway of the Cape Region appears to be the Cardon cactus. In order of decreasing dominance are Elephant Trees, Lomboy, and Palo Adan. Less dominant plants are Jumping Cholla (primarily in disturbed soils), Plumeria, Organ Pipe Cactus, Creosote Bush, and the yellow-flowered vine Yuca. Yuca is a morning glory relative covering plants like a large net.

The **CAPE PHYTOGEOGRAPHIC REGION** is outside of the Desert Region that covers most of the peninsula. The Cape Region, including the Cape mountains and part of the Sierra de la Giganta, has the highest rainfall on the peninsula. Most precipitation occurs during tropical summer storms.

The vegetation of the Cape Region is divided into the **Oak-Piñon Woodland** and the "impoverished" **Arid Tropical Forest**.

The **Oak-Piñon Woodland** community occurs in the granitic soils of Sierra la Victoria to an elevation over 6,000 feet (the transition life zone). The dominant species are Oaks, Black Oak and Pines, Palmita, Laurel Sumac, and Madrona. The **Arid Tropical Forest** of the Cape Region is an "impoverished tropical jungle". The trees of this forest are represented by the leguminous trees Palo Blanco and Palo Mauto, Mesquite, Acacia, Coral Tree, the edible Plum Tree; Palo Verde; and Plumeria with an understory of Palo de Arco, Pitaya Dulce, Lomboy, and Palo Adan.

	Vinorama *Acacia brandegeana*	Palo Blanco *Lysiloma candida*	Palo Mauto *Lysiloma divaricata*	Mesquite *Prosopis pubescens*
Bark	silvery-white	gray-brown smooth	thin, flaky	
Spine	slender straight single	legume without thorns	none	short, awl-shaped pair at node
Flower	yellow spikes Ball-like clusters	creamy white		yellow spikes
Pods	long, slender Copper-red at maturity	long, thin walled	twisted pods	tight spiral coiled pod screwbean
Range	flood plains Lower arroyos, mesas	below 600 m.	above 300 m.	below 1200 m. near water

BEAN TREES OF THE CAPE REGION

THE BEAN TREES OF THE CAPE REGION are an important part of the desert flora of Baja. Many of Baja's desert trees are legumes that produce a "bean" pod containing "beans" (seeds). Leguminous trees occur at low elevations in the desert, arroyos, and foothills below 3,000 feet. They are common in the Cape Region. The most commonly seen leguminous trees are Vinorama, Palo Mauto, Palo Blanco, and Mesquite.

PALO VERDE: Another non-leguminous tree commonly seen growing with the four leguminous tree species of the Cape Region is the Palo Verde in the Senna family. Species of the Palo Verde, a characteristic tree of the Sonoran

Desert north of the U.S./Baja border, grows all the way to the Cape region. *Cercidium praecox* is the Palo Verde of the Cape Region. It is distinguished from leguminous trees by its bright green photosynthetic trunk and branches (cladophylls) and leafless appearance. Palo Verde may be mistaken for a leguminous tree as it produces a pod-like fruit containing seeds the shape and size of pod beans. In the spring, it is covered with yellow flowers.

185   Turnoff to Todos Santos. Just south of the turnoff are large areas that have been cleared of native vegetation for grazing cattle.

THE PLUMERIA OF BAJA: The Plumeria is a common plant in this area. Plumerias grow as a shrub or tree up to 30 feet tall along canyons and foothill slopes, from just south of La Paz to Cabo San Lucas, in Baja's Cape Region. For most of the year, Plumeria is leafless, but is easily recognized by the showy long, white, tubular-flowers clustered at the ends of branches. The white blooms are abundant after summer rains. Plumeria are frequently seen in the gardens of ranchos where they are planted as ornamentals.

177.4   Over a small rise is a distant view of the high Sierra la Victoria. The road cuts are in pinkish alluvial material derived from the granitic basement.

Crested Caracara are occasionally seen on the ground. Several other birds, including the Red-tailed Hawk, White-winged Dove, and Aplomado Falcon are often seen sitting on the ground and perching on wires along the highway.

The Crested Caracara is a long-legged scavenger in the falcon family that spends much of its time on the ground. In flight, the larger head and beak, longer neck, white throat, and black and white-banded tail set it apart from the vultures. Its red-orange, bare facial skin is easily identifiable.

One of the best times to see Caracara is on the highway in the early morning, as it is the early bird that gets the road kill. Because of its long legs, they most often walk along the ground gathering food that consists of small creatures, carrion, and sometimes plant matter. They are a dominant bird often seen stealing food from other birds.

175   The highway passes over the Sierra Victoria Fault and into the main mass of the Sierra Victoria which has been uplifted along this fault. South of La Paz, the frontal fault of the Sierra Victoria cuts off the metamorphic rocks on the edge of the batholith. The highway crosses directly from the alluviated plain across the frontal fault into granitic and metamorphic rocks.

172.5   The highway passes through road cuts of pinkish granitic rocks.

167.4   Phyllitic schists are exposed here. The highway ascends through the foothills of Sierra la Victoria. These foothills are pocked with prospectors' holes. The hills are covered with Palo de Arco (*Tecoma stans*).

PALO DE ARCO is a small shrub or tree that produces large golden-yellow flowers after rains.

PALO DE ARCO

165   Altered and brecciated rocks are visible in this road cut. A fault zone separates the gneisses from the weathered granitic rocks.

164.5 The old mining region of El Triunfo.

163 The small mountain village of El Triunfo was established in 1862 with the discovery of silver and gold. This town was the center of a major silver mining operation after 1862. Two yellow church towers rise between the smoke stack and mine buildings on the hill. The ores are concentrated in dikes in the metamorphic rocks near the contact with the granitic rocks. A side trip to the main mines can be taken on the first road to the right after you enter town and cross the bridge. Follow this road for two blocks, go left at the wash, and cross an old brick bridge. The road wanders to the smelter and the old tower. Climb inside the tower and look up the old brick smoke stack to see either a small patch of sky or be treated to a view of the stars.

160   The highway passes through mixed gneissic and granitic rocks.

MIXED GNEISSIC AND GRANITIC ROCK

189

IMPOVERISHED RAINFOREST
LOMBOY – CARDON – PALO BLANCO – PITAYA DULCE

The hillside vegetation is primarily an **IMPOVERISHED RAIN FOREST** of several species of leguminous trees. Other plants seen are Pitaya Dulce (it looks almost like a Cardon in this environment), Elephant Tree, and Palo Adan. At this southern more mesic latitude, Palo Adan becomes less abundant and is largely restricted to the lower sandy Wash Woodlands.

158.8   There are granitic dikes in the road cut. The rocks at the crest of the pass are gneiss and dark-colored granitic rocks invaded by granitic dikes. The highway descends into San Antonio, another "old time" mining town.

156   The historical gold and silver mining town of San Antonio is located in the foothills of the Sierra la Victoria. This mining town was founded in 1756 when Gaspar Pison opened a silver mine over 100 years before the establishment of El Triunfo.

155.5   The highway ascends a grade with a view of Llanos San Juan de Los Planes, extending north to Punta Colorado and the Gulf. The volcanic and granitic Isla Cerralvo is offshore. The San Juan plains are part of the alluviated graben into which the highway descends after cresting the grade.

The vines entwined on many of the trees in this area are either the pink San Miguel (*Antigonon leptopus*) or the yellow Yuca (*Mersemia aurea*).

| YUCA | SAN MIGUEL VINE |

The **YUCA** vine is a member of the morning glory family which climbs upon and covers many of the plants of this area. When the Yuca die, they cause the host to appear as if it is covered by a net. Yuca have bright yellow trumpet shaped flowers that bloom all year, especially after a rain.

SAN MIGUEL VINE: The vine-like plant that grows on some of the trees of this region is known as San Miguel or Coral vine. This member of the buckwheat family reminds one of the commonly cultivated bougainvillea of southern California. Its seeds and potato-like tubers were eaten by Indians.

155   Gneisses exposed in the road cut. Good view of San Antonio.

154   This is the best view of the alluviated Llanos de San Juan region.

153   The highway passes through metamorphic rocks cut by granitic dikes while crossing the main horst and the granitic spine of the Sierra la Victoria. It then drops into a graben containing Santiago and Miraflores, then crosses a smaller horst to descend into the Gulf and San José del Cabo.

151   The Wash Woodland vegetation is Palo Adan, Elephant Tree, Jumping Cholla, Organ Pipe Cactus, Pitaya Agria, Acacia, Pitaya Dulce, and Lomboy. Cardons are growing in a narrow belt at the base of the hills.

Northern Mockingbirds are robin-sized birds with white breasts and black wings that show white patches in flight and a long narrow tail. Mockingbirds are expert mimics and repeat most of the songs of local birds while perched or in flight. These birds mimic more at night than do other members of the family Mimidae, an irritating fact if you are trying to sleep!

191

146.9   The highway splits in a wide "Y". To the right is San Antonio de la Sierra. To the left is Cabo San Lucas.

145.7   The granitic rocks at this kilometer mark have been dated at 73 million years. The flora is dominated by Mimosa and Pitaya Dulce. Palo Adan is seen as the highway climbs the slope.

Loggerhead Shrikes, seen throughout this region, are easily recognized by their heavy hooked beaks, broad black masks, and large white wing patches that contrast with their dark wings. Their head and back are bluish-gray and their underparts are white. They are known as  "butcher birds" as they habitually capture small lizards, insects, rodents, and small birds and impale them on cactus thorns, tree thorns, or barbed wire storeing their meal for later.

145   The forests of the Cape region are "double-canopy" forests. Palo Mauto and Acacia form a higher canopy that shades the lower plants; this produces a moister more mesic environment and reduces evaporation. The taller Organ Pipe Cactus grow at the bottom of the slopes where water collects as it drains off the surrounding hills.

143   Palo Mauto and Plumeria now occupy just the very high parts of the hills. *Acacia* begins to take over and becomes the dominant canopy tree.

135.5   After Rancho El Rodeo, the Candelabra Cactus is seen to the right of the highway. Cardon are mixed with the Organ Pipes along this stretch.

132   The highway begins to roughly parallel an arroyo vegetated primarily by Cheesebush.

130   There are exposures of light-colored granitic rocks with darker zenoliths along the road. A wild Fig tree is growing here and Yuca are becoming quite common. Broom Baccharis grow in the disturbed soils along the side of the highway with nearly leafless branches crowned by a green, stiff, broom-like mass. Often, bundles of branches are tied together and used as a broom.

128   Growing in profusion around the tropical agricultural village of San Bartolo are avocados, sugar cane, lemons, limes, mangos, papayas, figs, date palms, and fan palms. The vegetation of this arroyo is lush and tropical with *Acacia* as the predominating plant. Occasional Palo Adan, Elephant Tree, Palo Mauto, and Organ Pipes are scattered among the Acacia.

South of San Bartolo, the highway follows the northeast bank of the arroyo.

125   The stream terrace on the other side of the canyon was formed when sea level was higher and the streams graded the fans to that level. The highway will begin to follow this terrace at Km 121

123.5   Massive exposures of fresh granitic rocks are exposed just before Puente El Saldito.

120   This area was a basin at one time that in-filled with very coarse gray granitic gravels. Now this basin is being uplifted and dissected. These gray granitic rocks coupled with the sparse impoverished vegetation give the impression of a burned over area. However, this impoverished forest is lush and green during the summer rainy season.

119.5   The highway crosses the main wash of San Bartolo with outcrops of Pleistocene to sub-Recent sediments.

Mimosa (*Lysiloma sp.*), Tamarisk, and Lomboy are growing abundantly along the edge of the wash. Periodic floods, indicated by the rarity of trees, occur in this wash. The high energy of the floods is indicated by the width and depth of the wash. Cheesebush is the dominant shrub of the wash bottoms. It is able to reestablish itself more quickly than other plants; it will grow in the washes between the intermittent floods.

118   There are four levels of alluvial fan terraces that indicate different periods of uplift. The main older fan surface is on the far side of the yellow cliffs. Just below is a secondary surface that has large Palo Verde, Palo Mauto, and *Acacia* growing on it. A third level is closer to the highway with some smaller trees growing on it. And the fourth is the level of the wash bottom next to the highway. The higher hills beyond the fourth surface are alluvial gravels from a still older fan level.

112.8   The highway crosses Arroyo Buenos Aires with good exposures of alluvial fan material with large granitic blocks and sandstone lenses. These blocks are nearly all granitic with only a few metamorphic clasts.

Some parts of the hills to the southeast of Los Barriles are metavolcanic rocks. Moreno dated the metamorphic rocks on the west side of the Sierra at 73 million years which may represent a time of intrusion and metamorphism. The granitic rocks near El Triunfo have been dated at 73 million years and those nearer the Cape at 63 million years.

103   The highway passes through a relatively flat graben filled with at least 4500 feet of non-marine and marine sediments that have been uplifted and are now being dissected. The two sides of the graben can be seen in the long line of granitic hills to the southeast and in the edge of the granitic spine of the Sierra la Victoria to the west.

193

This thick series of Miocene to Recent nonmarine and marine basin-filling sedimentary rocks has been named: 1) the Coyote Redbeds, a Miocene alluvial fan deposit; 2) the Trinidad Formation, a middle to upper Miocene beach to outer shelf member and an upper Miocene to Pliocene slope and basinal deposit member; 3) the Salada Formation, an upper Pliocene shoaling upward sequence of outer shelf to beach deposits; and 4) a Pleistocene alluvial fan sequence (McCloy, 1984). Fossils are exposed in numerous arroyos in the basin.

92.5   Exposures of the marine Salada Formation with alternating fine and coarse-grained yellow sandstones are exposed in the road cut.

The road to the right leads to Las Cuevas (the caves) where there are Indian pictographs (paintings on rocks) *(See 5:118.3)*. The highway to the left leads 12 kms to La Ribera on the coast: then it goes 30 kilometers south along the Gulf to Cabo Pulmo, the only true coral reef in the Gulf of California.

Cabo Pulmo consists of a large bay with a fishing village. Skin diving is great along the reef with abundant tropical fish in a marine preserve. Just south of Cabo Pulmo is Punta Frailes, the eastern most point in Baja California.

CORAL REEFS are found where 1) the average water temperature is 78° and the coldest month is not below 60°, 2) there is average salinity, 3) there is a lack of turbidity, and 4) sunlight (less than 150 m deep). Corals can and do live in deep and cold water. These requirements are for the calcareous algae that bind the corals together to form the reef.

89   At the "crest" of the hill is a view of the Santiago graben. The Neogene formations stretch for kilometers to the base of the hills to the west. The fault at the base of the hills trends north-south, roughly parallel to the highway. The hills to the east form the east side of this graben. Caliente Manantials hot spring is located along the fault near Agua Caliente.

88   Excellent exposures of the turbidites of the Salada Formation can be seen along the highway for the next kilometer. They are massive sandstones with interbedded shales and siltstones.

84.7   Road to Santiago.

SANTIAGO GRABEN FILLED WITH TERTIARY SEDIMENTS

82.3   Crest over a small rise for a panoramic view of the graben. The straight nature of the fault of the Sierra la Victoria and the flat nature of the sedimentary rocks that filled the graben are obvious.

81.2   The Tropic of Cancer is the boundary between the Temperate Zone and the Tropic Zone.

70.9   The highway passes the turnoff to Miraflores. The craftsmen of Miraflores produce fine leather work. The road into Miraflores offers a view of the Sierra la Victoria. There are two high peaks on either side of a large pass almost straight ahead. The peak to the left is Cerro Picacho la Laguna, 6,207 feet. The peak to the right is 5,984 feet. These peaks show scars of the debris/mudflows that have stripped the vegetation along their path.

67   The highway drops into an arroyo with fine-grained fanglomerate sedimentary rocks on both sides and coarse fanglomerates above.

53   The vegetation is locally dominated by Elephant Tree, Jumping Cholla, Lomboy, Candelabra cactus, Palo Mauto, Palo Zorrillo, and scattered Acacias, Pitaya Dulce, Cardon, Palo Verde, Fairy-duster Mimosa, and Mistletoe.

43   Turnoff to Los Cabos International Airport.

32.5   The highway forks here. The right fork continues to Cabo San Lucas. The left fork goes into San José del Cabo. There are interesting orthogonal joints in the granitic rocks across the street from the Pemex Station. Since there are abundant granitic rocks in this area, the beautiful white beaches are all composed of quartz and feldspar.

195

24   The granitic rocks are overlain by fluvial sedimentary rocks (mostly decomposed granitic). These exposures continue for kilometers on the coast.

21.1   The highly weathered and jointed pink granitic rocks exposed in this road cut have been dated at 63 million years.

15   Many of the small offshore rocks and points along this coastline support small coral colonies.

6   There is a spectacular view of the cape, the bay, the arches, and stacks at the point of the Cape.

5   The fairly prominent terrace on the hill is also part of the main point. It reflects a higher stand of sea level at 10 meters. There is a large coastal dune field at about the 10-meter terrace level that extends up the coast for approximately five kilometers.

2   The highway crosses a wash. Take the road to the right to Todos Santos. To continue all the way to the cape, follow the divided highway around the harbor.

FRIAR'S STACKS OR ARCHES: The arched rocks visible at the end of the cape of Cabo San Lucas are known as "The Friars." The Friars are stacks, isolated rocky islands detached from the tip of the peninsula by wave erosion. At some time in the future, The Friars will be completely destroyed by wave erosion along joints and differential erosion of the weaker sections of rock. Wave action may hollow out cavities or sea caves in the cliff, and if this erosion should cut through, a sea arch is formed. The collapse of a roof of a sea arch leaves an isolated mass of rock called a stack. This area is commonly referred to as "Land's End."

AT CABO SAN LUCAS - A short hike on the Pacific beach, over the low headland, will lead to Land's End and a beach to cross the Peninsula.

CABO SAN LUCAS FROM THE HIGHWAY

CABO SAN LUCAS FROM THE BEACH SAND RIDGE. AFTER I TOOK
THIS PHOTO I WAS TREATED TO A FULL ECLIPSE OF THE MOON
WHILE LYING IN MY SLEEPING BAG.

XENOLITH IN GRANITICS – With a little imagination I can make out the upper legs, knees and, lower legs, body, long neck, head, eyes, and nose

The granitic rocks between Hotels Solomar and Finisterra are highly jointed and contain large **xenoliths**.

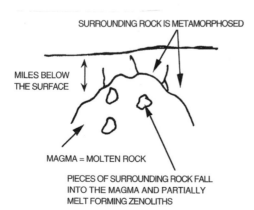

SURROUNDING ROCK IS METAMORPHOSED

MILES BELOW
THE SURFACE

MAGMA = MOLTEN ROCK

PIECES OF SURROUNDING ROCK FALL
INTO THE MAGMA AND PARTIALLY
MELT FORMING ZENOLITHS

XENOLITHS: With the intrusion of magma, the molten rock melts and pushes its way into the surrounding rocks. Pieces of the surrounding rock are broken off and sink into the magma. Some of these pieces are partially melted and the resulting piece is generally richer in iron and thus darker than the igneous rocks of magma. These pieces are called xenoliths. Miles of uplift and erosion raises everything to the surface.

198

VIEW FROM GULF BEACH AT LANDS END

LAST RAYS OF THE SETTING SUN ON A CABO PINNACLE

SURF ON THE LAST PACIFIC BEACH AT LANDS END

Log 10 - Cabo San Lucas to La Paz via Todos Santos [157 kms = 97 miles]

From Cabo San Lucas, the highway heads northward traveling largely on the dissected surfaces of alluvial fans from the granitic Sierra la Laguna. Extensive dune fields and supra-tidal flats are near the road where it lies close to the ocean. Near Todos Santos, the road passes through several low passes in the granitic and metamorphic rocks. At Todos Santos, the road turns inland to travel on a dissected alluvial surface along a major fault close to the granitic hills of the Sierra la Laguna.

Note: There have been major revisions of the highway route from development to realignment and widening of the road to four lanes. The revisions to this log are subject to some variation; however, most of the kilometer points are correct or close and the geology is good. Some of the route descriptions may be a bit off.

123   There are two routes from Cabo San Lucas along Highway 19 to Todos Santos and La Paz. Retrace the route back to the Pemex station and turn left on Highway 19 north or take the route from town on Paseo Morelos. The route is somewhat confusing as there are many new roads.

From Cabo San Lucas, the highway goes northwest on alluvial fans toward the granitic Sierra la Laguna and passes through the Cape phytogeographic region. The vegetation along the highway consists of a heavy thorn forest underbrush dominated by Organ Pipe Cactus, Cardons, Elephant Tree, Yuca, Lomboy, Palo Mauto, and Cholla.

118   Junction with the Toll Road to San José de Cabo.

115   The highway passes through an area of mature topography and alluvial slopes. The lack of outcrops in this area is the result of the deep tropical weathering of the granitic rocks and subsequent burial of the outcrops by alluvium. The present streams are cutting into the fans to establish another surface a few 10's of feet lower than the present surface.

113   Organ Pipe Cactus with a sparse desert "understory" of Lomboy become the predominant plants in the cape area.

The highway passes through several road cuts in the highly weathered granitic rocks. Somewhat lesser weathered granitic rocks are visible in the washes of this region.

ORGAN PIPE CACTUS WITH A "UNDERSTORY" OF LOMBOY

111   The highway crests a pass with sweeping views of the Pacific Ocean to descend a long grade on the alluvial fan surface toward the Pacific.

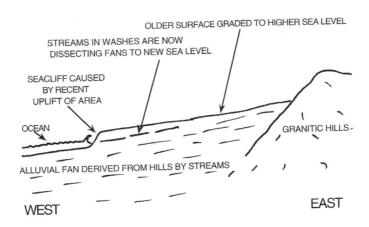

104   As the highway crosses a wash, excellent exposures of granodiorite can be seen in the lower part of the wash.

103   The species composition of the vegetation is essentially the same as that listed at Kilometer 123, but the trees are noticeably shorter, rarely exceeding 9-12 feet in height. Lomboy and Leatherplant dominate the

understory of a short "impoverished forest" of Elephant Trees, Cardon, and a few scattered *Acacias*.

Stanch it with **LOMBOY OR NATURE'S "CHAP STICK":** The low shrubby Lomboy (*Jatropha sp.*) seen along the highway in this area is a member of the spurge family. Members of this family usually produce a milky acrid sap.

LOMBOY

Over 8,000 species of euphorbs grow around the world. Some furnish food and valuable oils while others are of ornamental or medicinal value. The two species of *Jatropha* commonly seen along Baja's highways on plains, hillsides, mesas, and sierras are *J. cinerea* and *J. vernicosa*. The sap of *J. cinerea* is highly astringent and is said to stanch bleeding wounds, prevent chapped lips, and, unfortunately, permanently stain clothing.

ELEPHANT TREES of this area are represented by the unrelated species of the two genera Pachycormus and Bursera. Because they store water in the cortical cells of their elephantine trunks, Elephant trees grow luxuriously despite the extreme aridity of this region. The Bursera are easily distinguished by the incense-like odor produced by crushed leaves. *Pachycormus* tissues are odorless.

3

PALO ADÁN is a relative of the Ocotillo which is commonly seen in the northern deserts of Baja. Palo Adan has thicker branches, a trunk, smaller flowers, and normally grows from Parallel 28° south to the Cape Region on the clay and granitic soils of alluvial plains.

100.5 Granitic rocks cut by dikes are exposed in the road cuts.

98   Here is another view of the Pacific Ocean. Numerous active and stabilized dunes near the coast are to the west. For the next few kilometers, the highway parallels the coastline that ranges from 1 to 2 kilometers to the west of the highway. Along this stretch, there are almost continual views of the Pacific Ocean and its fringe of active and stabilized sand dunes.

The vegetation in this region has become sparse. The taller plants are Elephant Trees and Cardon, and shrubby, leafless Lomboy and Leather Plant dominate the "understory." The highway drops into another wash where the dirt road to the west leads approximately one kilometer down the wash toward the Pacific and ends at a beautiful sandy beach with rocky points and headlands composed of granitic rocks.

84   Beach.

82   From this point, an alluviated surface stretches to the beach. It is truncated at the 10 meter terrace level above sandy beaches. A beautiful beach is located at the bottom of this gentle grade near Las Cabrillas. However, it is unsafe to swim here as the surf in this area is dangerous.

78.5   View of the ocean and several ranchos that are surrounded by Fan Palms and fruit trees.

76.5   A short walk in this area makes possible a close inspection of the coastal storm beach berm, supper-tidal flats, and alluvial surface.

75.5   The highway crests a low pass through a major road cut in gneiss. The view to the north is of several kilometers of the Pacific coastline beaches.

70   Another good view of a supper-tidal flat (Salitrales). Salitrales are often dry a good part of the year. Farther north on the west coast of Baja they form hard surfaces that are excellent to travel upon. However, when they flood during the summer rainy season or very high tides, they become slick and "soupy" and not usable.

61   The highway traverses the alluviated fan surface with views of Punta Lobos. Gneisses and other metamorphic rocks form Punta Lobos.

57   Bypass cut off.

52.2   Todos Santos is an agricultural village developed in a green valley near the Pacific and close to the Tropic of Cancer. The town began as a farming community and a mission visiting station in the early 1700s.The tall trees that line the highway into Todos Santos are Mango Trees.

47.5   Bypass road rejoins the old road.

45   View of the foothills of the northern Sierra la Victoria. The frontal fault of the Sierra la Victoria cuts diagonally toward the highway from right to left.

39   The highway descends into a large arroyo through granitic rocks. A thin terrace covers the granitic rocks while coarse blocks of granitic debris are visible downstream.

North of Todos Santos, the highway roughly parallels the Sierra la Victoria Fault Zone that continues across Baja Peninsula and passes east of La Paz. To the left is the faulted graben of Llanos de Santo Tomás. The Todos Santos Graben is underlain successively by Pliocene sedimentary rocks, Miocene volcanics, and marine sedimentary rocks, and ultimately by a Cretaceous syncline that is truncated by the Sierra la Victoria Fault.

Locally, the Sierra la Victoria has been planed off into a pedimented surface.

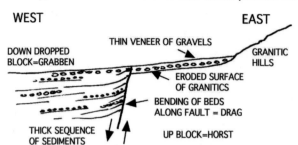

WEST

THIN VENEER OF GRAVELS

DOWN DROPPED
BLOCK=GRABBEN

EAST

GRANITIC
HILLS

ERODED SURFACE
OF GRANITICS

BENDING OF BEDS
ALONG FAULT = DRAG

THICK SEQUENCE
OF SEDIMENTS

UP BLOCK=HORST

The pediment is a sloping erosion surface developed at the base of the abrupt and receding front of the Sierra la Victoria. The pediment is underlain by granitic and metamorphic bedrock mantled with a thin discontinuous veneer of alluvium derived from the Sierra la Victoria. This thin veneer passes over the fault and becomes quite thick on the down-dropped side of the fault. There are a complex series of faults in this frontal fault zone.

An unconformity consisting of gneiss overlain by the alluvial gravels is exposed at Km 4 on the bypass.

36   Slates in road cut.

29   There is a view to the rear of the Sierra la Victoria. The highway continues to the northeast across the flat alluviated surface and then turns to the east toward La Paz. As the highway traverses this flat featureless plain, it approaches the Sierra Las Calabazas. This small range, resembling bald heads, is visible to the right of the junction of Highway 19 and Highway 1.

Keep your eye out as you may see a Roadrunner (Cuckoo Bird) (*Geococcyx californianus*) run across the road on this section of the highway.

The **Greater Roadrunner** is the large terrestrial Cuckoo Bird seen running on the ground throughout the entire peninsula. The most identifiable field characteristics of the roadrunner are its bushy crest, brown and white streaked back, and its long, black, white-tipped tail. A closer look at the eye will reveal that it's flanked by beautiful red, white, and blue feathers. The food of this fleet-footed predator consists of anything that moves (primarily lizards, snakes, and insects). This bird's scientific name translates to "California earth tail" referring to its running habit and long tail.

0   Junction. Go north 34 kms. to La Paz or south to Cabo San Lucas.

Log 11 - Laguna Chapala to San Felipe via Gonzaga. [215 kms = 133 miles]

Note: This highway was still under construction in the spring of 2017. The new route is high on the side of the rugged metamorphic and granitic hills as they avoid the major canyon that the dirt road followed. The construction crews appeared to be in the initial stages of making major cuts along this new alignment. This log covers the first few kilometers from Laguna Chapala and starts again about 17 kilometers south of Bahía Gonzaga. There is approximately 30 kilometers of rough unlogged dirt road between these points that will soon be abandoned. It is anticipated that the kilometer distances south of Bahía Gonzaga will change once the new, shorter road is completed.

0   The highway proceeds east, across Laguna Seca Chapala.

2.6   This is the Old Rancho Laguna Seca Chapala which is now abandoned. The site was chosen because of the mines that are low on the hill of granitic rock to the north, and because the old main road down the peninsula intersected another crude road going east to the mines at Las Arrastras and to Bahía San Luis Gonzaga. For the next 5 kilometers, the highway will travel through a granitic and metamorphic terrain.

7.7   After crossing a small fault, the highway climbs out of the arroyo over a hill in a belt of metavolcanic rocks that have been faulted between granite, slate, and schist.

8.6   The highway crosses first into granitic rocks, then into metamorphic rocks, then makes a bend up a canyon in the metamorphic rocks.

The vegetation consists of Creosote, Brittlebush, Garambullo, Cardon (small and sparse), Cirio, Deadly Nightshade (opportunistic along the road), Ocotillo, and shrubs such as Rabbit Brush and Purple Sage.

At this point, traffic was diverted to the old dirt road down a canyon due to the major construction of the deep road cuts. The log will pick up, after 30 kilometers, on the alluvial fan at kilometer 164, approximately 17 kilometers south of Gonzaga Bay.

If you want to do some off highway exploring, find the road to Calamajué Canyon. A trip up this wide canyon bottom will lead to the Calamajué Soda Springs.

The old Gulf road down Calamajué Canyon. *The sandy wash was aggrading [slightly filling] resulting in a smooth [fairly high speed] road with few cut-banks to cross.*

Calamajué Soda Spring *- The water is carbonated and not bad. Makes a good mixer.*

CALAMAJUÉ SODA SPRINGS - Tequila + soda water + pickle juice - On the road north to San Felipe we usually drove down Calamajué Canyon and stoped at Calamajué Soda Springs. We frequently stopped here on our way north to the U.S. The springs are carbonated and make a good mixer. We were usually out of everything except the trading Tequila and pickle juice at this point in our travels. We affectionately called our concoction a **"Calamajué sling." Don't knock it until you try it.**

The next morning, we would walk over to the outcrop and knock off a bi-carb. The dissolved solids in the spring water formed the terrace you see here. Can you see the face? It is winking at you.

BACK TO THE MAIN ROAD SOUTH OF BAHIA GONZAGA

164   The highway travels on the high, gravelly Quaternary terrace with stands of Ocotillo and tall Elephant Trees with sand Verbena along the roadside.

The Quaternary fluvial terrace material that the road is following parallels the present-day San Francisquito Creek. The terrace probably represents part of the ancestral bed of the creek when sea level was higher. When sea level lowered, this area was abandoned by the creek, leaving a relatively smooth terrace. The creek is vegetated by a vast expanse of Ocotillo, Brittlebush, Burroweed, Creosote Bush, and scattered Teddy Bear Cholla.

ALLUVIAL FAN BORDERED BY A WASH AND BOULDERY TONALITE HILLS CAPPED BY VOLCANIC FLOWS

Think of the history that this photo displays. The tonalite crystallized from molten rock formed in a magma chamber miles below the surface during the early Mesozoic era. The area was then uplifted and miles of overlying rock were stripped away by weathering and erosion to expose the tonalite and form a relatively flat erosion surface. This surface was relatively undisturbed until the East-Pacific Rise began to rip Baja from the rest of the North American Continent in the Miocene. Lava from this breakup poured over the erosion surface leaving a thin cap of volcanics. (There is some 50 to 60 million years missing between the formation of the tonalite and the volcanic rocks.) Tilting of the Baja peninsula, faulting, uplift, weathering, and erosion along the Gulf edge of Baja produced the hills. *The Story doesn't end here.* The present alluvial fan surface was deposited in the late Pleistocene (about 125,000 years ago) during a high stand of sea level. The wash along the edge of the hills is currently eroding the fan surface to a lower level in response to the present lower stand of sea level.

159   First view of Ensenada San Francisquito and Bahía San Luis Gonzaga. The low hill of the far point directly ahead is Isla San Luis Gonzaga. The small point to the left is Punta Willard the north point of Bahía San Luis Gonzaga. The island and Punta Arena separate the main part of Bahía San Luis Gonzaga from the Ensenada San Francisquito.

SAN FELIPE DESERT FLORA: CORDON, CIRIO, OCOTILLO, CHOLLA, CREOSOTE, CHEESEBUSH, BURROBUSH, YUCCA, AND LOW ANNUALS ON THE ALLUVIAL FAN BACKED BY METAMORPHIC AND GRANITIC (TONALITE) HILLS

156   Turnoff to Punta Final.

153   The vista ahead reveals good views of Isla San Luis Gonzaga, Alfonsina's airstrip on Punta Arena, Bahía San Luis Gonzaga, and Ensenada San Francisquito. In the far distance, the gray point is Isla San Luis.

The highway passes through a wide part of the Arroyo las Arrastras. The stream has incised about 10 meters below the level of the braided stream drainage of the old alluvial fan. The upper surface probably represents the surface graded to the Sagamonian 5E high stand of sea level about 125,000 years ago.

147.2   Turnoff leads to Alfonsina's airstrip.

144   The highway passes through a roadcut blasted into andesite material. The ground here is littered with dark colored cinders.

141   The highway rounds a corner and drops slightly into a flat alluviated area. The Pliocene marine sedimentary rocks seen here form a "badlands-like" topography under the alluvium. The highway is flanked by andesite hills.

140   The range to the left is all rhyolite. This roadcut is in the altered sedimentary rocks of the Pliocene Salada Formation.

PLIOCENE MARINE SEDIMENTARY ROCKS

At the crest, the vista opens with the gulf and a number of volcanic islands

On the alluviated plain, the reddish and yellow beds of the Pliocene marine sedimentary rocks of the Salada Formation are exposed to the left in the gully below. They are overlain by terrace material. The bouldery outcrops on the higher hills behind the Pliocene are spheroidally weathered tonalite. The andesite volcanics drape over the older formations. The highest hill is a small basalt shield cone called Cerro el Portrero.

The highway descends through the Pliocene sedimentary rocks of the Salada Formation onto the flats, with very low hills underlain by Pliocene rocks in the near distance ahead.

The vegetation covering the Pliocene marine terrace consists of Ocotillo, Palo Adan, and Creosote. It is sparser here than it was at the vista point back at the crest of the hill. Smaller Smoke Trees, Rabbitbrush, and Burroweed predominate in the washes along this stretch of the highway.

138.7   Pliocene beds are exposed in the roadcut at the edge of this wash. The low hills to the left exhibit exposures of the Pliocene Salada Formation. Volcanics are visible to the right. Along this stretch, the highway alternately climbs on the low fluvial terrace, drops into a wash, and back to the terrace.

From Bahía San Luis Gonzaga north to San Felipe, the highway passes through the San Felipe Desert subdivision of the Desert Phytogeographic Region (*See* 14:108).

The following bird species are typical of the San Felipe Desert:

BIRD NAME	LIKELY LOCATION
Abert's Towhee	Desert woodlands and streamside thickets
American Kestrel	Wires and fence posts
Amer. White Pelican	Gliding along the shore
Anna's Hummingbird	Red or yellow tubular flowers
Bendire's Thrasher	Flies from bush to bush, feeds on ground
Cactus Wren	On cacti
Calif. Brown Pelican	Gliding along the shore
California Quail	On the ground
California Thrasher	On the ground
Costa's Hummingbird	Feeding on red flowers
Crissal Thrasher	Secretive, hiding in brush
Gambel's Quail	On ground in desert scrublands and thickets
Gila Woodpecker	Nests in holes in giant cacti
Greater Roadrunner	Crossing the highway
Ladder-back Woodpecker	Within the desert, sometimes nest among the agave stalk
LeConte's Thrasher	In sparse vegetation, flies when necessary
Loggerhead Shrike	Wires and fence posts
Red-Tailed Hawk	Tops of telephone poles and fence posts
Scrub Jay	In chaparral and woodlands
Turkey Vultures	Soaring in the skies or feeding on carrion
Vermilion Flycatcher	Streamside shrubs and wooded ponds
Western Meadowlark	Fence posts and fence wires

The vegetation has dramatically changed. Except for the Creosote Bush and several species of annual grasses, almost all of the plants have disappeared. A few very small Ocotillo can be seen growing on the rises between the washes.

136   The north end of Isla San Luis is directly east of the highway.

133.5 Turnoff at Punta Bufeo. At the north end of Punta Bufeo, the vegetation consists entirely of Creosote Bush, few Brittlebush, annual grasses, and Ocotillo that appear mostly on the edges of the washes or scattered about on the Fan surface.

130   The view northward across the alluviated fan is of Isla San Luis and most of the other small offshore islands located to the north of Punta Bufeo.

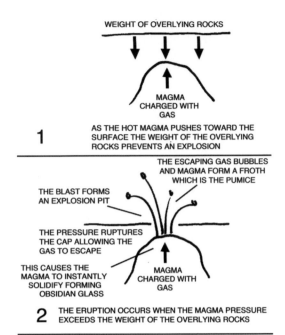

WEIGHT OF OVERLYING ROCKS

MAGMA CHARGED WITH GAS

1 AS THE HOT MAGMA PUSHES TOWARD THE SURFACE THE WEIGHT OF THE OVERLYING ROCKS PREVENTS AN EXPLOSION

THE ESCAPING GAS BUBBLES AND MAGMA FORM A FROTH WHICH IS THE PUMICE

THE BLAST FORMS AN EXPLOSION PIT

THE PRESSURE RUPTURES THE CAP ALLOWING THE GAS TO ESCAPE

THIS CAUSES THE MAGMA TO INSTANTLY SOLIDIFY FORMING OBSIDIAN GLASS

MAGMA CHARGED WITH GAS

2 THE ERUPTION OCCURS WHEN THE MAGMA PRESSURE EXCEEDS THE WEIGHT OF THE OVERLYING ROCKS

OBSIDIAN DOME

PUMICE RING

PUMICE BLANKET

THE PUMICE IS BLOWN OUT TO FORM A BLANKET OVER THE AREA AND A RING AROUND THE EXPLOSION PIT

OBSIDIAN

3 FURTHER UPWARD MOVEMENT OF THE MAGMA FORMS THE OBSIDIAN DOME

ISLA SAN LUIS consists of several low rings of pumice (the gray material) with very obvious obsidian domes (the darker material). The same geological conditions exist at Mono Craters in central California. The south end of Isla San Luis is an explosion pit consisting of a large pumice ring with no dome. The southern point of the island is the west rim of the explosion pit. The island's cliffs were formed by the sea eroding the soft pumice. The obsidian on the larger dome was dated using obsidian hydration rates and found to be as young as 100 years. The darker material at the north end of the island is another obsidian dome.

The tonalite hills ahead are covered with the brown mottling effects of desert varnish. The light-colored material cutting the hills in streaks are dikes.

128 The highway crosses the Arroyo Mal de Orin. The prominent wash woodland vegetation of this arroyo consists of Smoke Tree, Elephant Tree, Horehound, Princess Plume, Spanish Bayonette, Yucca, Pencil Cholla, and mistletoe laden Palo Verde.

126 Garambullo and Cheesebush are the tall floral dominants. An interesting member of the squash family, a bush with a yellow-orange, trumpet like flower called a Devil's Claw or Unicorn Plant can be sighted. The highway has been traversing an active alluvial fan for several kilometers.

118 The vegetation has changed dramatically. The ground is almost entirely obscured by a thick cover of Brittlebush. In the springtime, the area is illuminated by the bright yellow flowers of the Brittlebush. There are a few scattered Creosote whose numbers increase toward the beach. The highway

approaches the end of the coastal alluviated Pliocene marine terrace. The low hills ahead are volcanic. The tonalite on the left is cut by numerous nearly vertical, light-colored, ribbon-like dikes. A foothill "screen" of dark metamorphic rocks is exposed close to the volcanic hills forming a low belt of darker colored, steeper hills than the volcanic hills ahead.

VIEW TO SOUTH OF COAST AND ISLANDS

BURSAGE

The Bursage, seemingly the only plant growing here, looks like little gray "buttons" all over the hills. Because each plant has its own nutrient and water requirements, they space themselves naturally, appearing to have been planted in rows. At lower elevations between the highway and the gulf, Creosote and Ocotillo are occasionally growing among the Brittlebush. Atriplex and Wild Buckwheat are also seen scattered among the Brittlebush.

113   The highway passes through a series of road cuts in volcanic rocks.

Magnificent Frigate birds are commonly seen soaring off shore (7:94.3).

104.   View of the village of El Huerfanito and the white, bird guano covered Isla El Huerfanito. To the left of El Huerfanito is a view of the highway

climbing up the first of the steep grades of the Cuesta La Virgin. The dark-colored hill midway up the coast is the basalt cone, Volcan Prieto.

102   Climbing the grade provides a view of a series of stripped dip-slopes, dissected by streams, reaching to the ocean. Below the slopes the streams valleys form steep-sided, fjord-like embayments.

98   Near the crest of the first big grade is a pinkish rhyolite with a 15 to 30 cm. baked zone. Below the baked zone is a very muddy rhyolitic or andesitic lahar that looks like a mudflow and a layer of basaltic cinders.

The old highway eventually dropped into a major wash (Arroyo Heme), the first north of the first grade. The highway then climbed out of the wash to a view of a nice rocky beach with a storm berm. The new highway remains fairly level with new road cuts on the ridge. There are views of the old road in the canyon.

In places, the view of the volcanic tableland looks deceptively flat, as you are looking at the dip-slope. Where this slope is cut by canyons, it is revealed as being a steep slope.

91   Along the right side of the highway at the very southern end of the large basaltic cone, there is a large sandy tidal flat, often below sea level. Fresh water from the arroyo or sea water can fill this entire area during high tides and storms.

85 The hills nearer the highway are covered with white *Eriaganum*, another species of wild Buckwheat. The other hills are covered with Brittlebush, Ocotillo, and Mistletoe laden trees.

84 There is a road going down to a small settlement on a fine gravelly beach. A tómbolo extends to a small, near shore rocky island at low tide. A dip-slope has created a barrier with many excellent, beautiful tide pools.

77.5 The grade just south of Puertecitos provides a nice view of the harbor. Look for Magnificent Frigates, Brown Pelicans, and Gulls in the bay.

BURROS - BURSAGE - RHYOLITE

74.3 **PUERTECITOS** is a northwest-southeast oriented, linear bay formed by faults along both sides of the bay. The rocky shores (horsts) on the west and east are formed by uplift along two faults with the down-dropped (graben) block forming the sandy beach in the middle. Due to an 8 meter tidal range in this area, the high tide on this beach will come up to the foundations of the buildings. At low tide, the bay is almost devoid of water. Toward the edge of the point, about 3/4 of the way down the east side of the bay, there is a hot spring located along the fault. At low tide, the steam can be seen rising from the area. The highway closely follows the fault line on the west side of the graben. The cliff on the left side is a fault scarp. The fossiliferous Pliocene beds are exposed near the airport. The highway climbs out of the graben through a low pass in the hills. Most of the dark rock on the beach side of the highway is rhyolite.

PUERTECITOS AT HIGH TIDE

PUERTECITOS AT LOW TIDE

74   Beautiful rocky points and pocket coves are present along the highway where the resistant rhyolite beds dip into the ocean.

68   North of Puertecitos, the highway drops into the large Arroyo Matomi and runs "as straight as an arrow" across alluvial fan material.

61   The low hill on the horizon to the southwest (left rear) is Picacho Canelo, a rhyolitic volcanic eruptive center.

58   Toward the end of the plain, the highway climbs on the terrace in the foothills of the Sierra San Fermin, makes a deviation to the left, and crosses more washes and terraces.

54.5   The side road to the right leads down to Playa Cristina with a small lagoon and a coastal dune field.

53   The highway drops from the elevated terrace back onto the alluvial fan and crosses Arroyo de Chale.

37   The vegetation changes to a Creosote Bush/Ocotillo community typical of the San Felipe Desert *(See 14:108).*

30.6   Bahía Santa María. The road to the left goes to Agua de Chale and the sulphur mines. The sulphur was brought to the surface by vapor and hot water along a fault line. The hot spring or fumarole activity has altered the rhyolite beds and deposited the sulfur as small yellow crystals and crusts in the cracks and cavities in the rocks.

The coastline here has a wide mudflat in back of the beach behind the storm berm. The high hill to the north between Punta Estrella and Punta Digs is Cerro Punta Estrella, a largely tonalite hill with some prebatholithic metamorphic rocks along its flanks.

The vegetation stabilizing the coastal dunes is primarily Creosote Bush, Mesquite, Atriplex, Smoke Trees (in the washes), Garambullo, Palo Adan, Brittlebush, and *Acacia*. The flats contain relatively low vegetation.

20.5   Road to super-tidal flat of Laguna Percubu.

18   The tonalite pluton of Cerro Punta Estrella straight ahead is becoming impressive. It looks as though the highway is heading straight into the mountain range. In this area, everything seems to be at a great distance.

5   The highway drops behind and follows along the edge of the 30 to 45 feet high semi-stabilized dune.

The Sierra San Pedro Mártir and Picacho del Diablo, the highest peak in Baja California (3,115 meters, 10,126 feet) are in view on the left front.

The dunes in this region are moving. Some of the vegetation growing on or near these dunes consists of Ephedra, Creosote Bush, Ocotillo, Atriplex, and Cat's Claw Acacia. These last two plants are deep-rooted and the dunes hold moisture so they can survive in this area. On hummocks, the sand is held in place by the roots. As the sand builds up, the tree grows higher.

219

0   The highway divides. The road to the left goes to the airport; the one to the right climbs on the dunes and goes about 5 Kms. into San Felipe.

As the highway crests over a dune, the beautiful expanse of Bahía de San Felipe and the town of San Felipe comes into view. The coastline here tends to be muddy because of the Colorado River and the volcanic terrain that produces finer-grained material. As you drive down the highway along the crest of this stabilized dune, you can see the sand being blown across the highway because there is little brush to stop it.

The highway soon crosses an arroyo. The right point of the harbor is Punta El Machorro consisting of tonalite. The slightly higher hill to the left is Cerro El Machorro, consisting of granodiorite. The low-lying hills in the foreground, behind the city, are pre-batholithic carbonaceous rocks of undetermined age.

Where to from here: Follow the reverse of Log 14 to Mexicali or at Km 140 turn west and follow Log 12 to Ensenada.

REMOTE AREA OF MOUNTAINS AND BASINS

Log 12 - San Felipe to Ensenada via Valle Trinidad [245 kms = 152 miles]

San Felipe to Baja Highway 3 - *The highway climbs onto uplifted alluvial fans and passes through a series of steep granitic and metamorphic hills. For several kilometers north of the hills, the highway undulates across the dissected uplifted fans. The granitic Sierra San Pedro Mártir form the high mountains on the far left. The largely granitic Sierra San Felipe form the nearer desert varnished foothills. As the highway crosses the alluvial fans, the last views of the Gulf of California fade into the distance across the super-tidal flats of the Salinas de Omotepec. The highway crosses several dune areas with sand from the Gulf and supper-tidal flats. The highway continues for many kilometers along the alluvial fans descending from the rugged Sierra de San Felipe with views ahead of the rugged Sierra Pintas.*

Baja Highway 3 to Valle Trinidad - *From the intersection of Highway 3, the highway heads west across uplifted dissected alluvial fans, past the rugged hills of granitic and metamorphic rocks of the Cerro El Borrego and the Sierra de San Felipe, towards the crest of the granitic and metamorphic Sierra San Pedro Mártir. After skirting the north end of the alluviated Valle San Felipe graben, the highway enters a rugged, steep area of granitic and metamorphic rocks in San Matias Pass at the end of the Agua Blanca Fault Zone. The highway then follows, at grade, the trace of the Agua Blanca Fault Zone through steep hills of granitic and metamorphic rocks to the broad alluviated Valle Trinidad. Miocene volcanic rocks cap the mesas to the north.*

Valle Trinidad to San Salvador - *At Valle Trinidad, the highway turns and climbs a grade on a scarp of the Agua Blanca Fault, through rugged hills in the granitic and metamorphic rocks onto the flat El Rodeo surface. Miocene volcanic rocks form the mesas to the northeast. The highway travels through rolling hills in the granitic and metamorphic rocks, then skirts a large flat alluviated area and approaches a low line of hills that mark the approximate trace of the San Miguel Fault Zone. The highway then descends a gentle valley in the tonalite hills along the San Miguel Fault Zone to San Salvador.*

San Salvador to Ensenada - *The highway follows a gentle valley along the San Miguel Fault Zone in the rugged granitic and metamorphic hills, then turns west and descends a relatively smooth rocky surface to the broad flat alluviated Ojos Negros Valley. It then passes one of the numerous isolated steep metamorphic hills that dot the valley, and heads toward an escarpment of rugged gneiss hills. The highway crosses the frontal fault and passes through rugged bouldery granodiorite hills to descend a steep valley in the rugged hills of granitic and metamorphic rocks and climb a steep grade in rugged metasedimentary and metavolcanic hills. After going over a pass, the highway descends the narrow, rugged Arroyo del Gallo in the metavolcanic, gabbro, and tonalite hills. After climbing out of the arroyo at Piedras Gordas, the road follows a rolling ridge through tonalites, gabbros, and gneisses and finally descends a grade through rolling tonalite slopes to Ensenada.*

Follow the reverse of Log 14 to the San Felipe Junction. If you are returning to Mexicali, follow the reverse of Log 14 to Mexicali.

140   Junction of Highway 3 to Ensenada (195 kilometers) and Tijuana (300 kilometers). Turn to the west. The scarp of the Sierra San Pedro Mártir Picacho del Diablo are directly ahead. The Sierra Borrego are to the north; the Sierra San Felipe are to the south. The Sierra Pintas are to the far north.

195   The kilometer markers now decrease to Ensenada. The crest of the Sierra San Pedro Mártir (except for the peak of Picacho del Diablo) is very flat. This face of the range is actually a 10,000 foot high scarp. Valle San Felipe in front is very close to sea level. The valley has over 6,000 feet of alluvial fill, making the offset on the fault in excess of 16,000 feet (3 miles).

194   The highway makes several bends and heads directly toward Cerro El Borrego, the high peak of the Sierra San Felipe, a granodiorite pluton (slightly to the right of the highway). The lower grayer rocks around it are pre-batholithic carbonate sedimentary rocks. The lower hill to the left of Cerro El Borrego is also granodiorite. To the north, the hills are a mixture of granodiorite, tonalite, and pre-batholithic carbonates until, at about 3 o'clock, they become the Miocene volcanic rocks of the Sierra Pinta. The vegetation is largely the same. As in other parts of Baja, sometimes it seems like it is all Ocotillo, other times all Smoke tree or all Mesquite; however, it is just dominance that is changing. In the spring, this section of desert is often ablaze with the yellow flowers of the Brittlebush.

186   The highway follows an elevated fan with washes on either side that are about 30-60 feet below the fan. This whole area has been uplifted slightly and the washes are now regrading the fans to the new base level. Here and there the road drops down into one of these washes where the Mesquite and Smoke Trees become dominant. The little gray rounded balls are Burro Brush and Brittlebush which are more plentiful in the washes than up on the fans. The bright green plant along the road is Cheesebush.

179   Pass abreast of the Sierra San Felipe. There is a granodiorite hill to the north with a low skirt of metamorphic rocks around it. Notice the difference between the weathering characteristics of the Sierra San Felipe, which is very light, and the darker granodiorite cut by a number of small dikes that form the low hills at about 2 o'clock.

173   The highway begins to pass by and through a series of hills to the left that are part of the old Pliocene alluvial fan that has been largely removed by the present drainage, leaving isolated hills. In the Pliocene era, the top of these hills would have been the surface of the alluvial fan. Now, the surface has been elevated and forms hills that are being dissected by the new stream drainages. In this desert terrain, you typically get severe erosion

of older sedimentary rocks that are then remixed into the newer sediments and washed down into the washes.

CHEESEBUSH   CHOLLA

PALO VERDE   OCOTILLO

171   Roadcut in Pliocene conglomeratic sedimentary rocks. Out of the valley into the higher elevation, there is more Ocotillo, Spanish Bayonette, Teddy Bear Cholla, Creosote (replacing Cheesebush as a roadside opportunist), Barrel Cactus, and Brittlebush. Teddy Bear Cholla and Creosote form a thick underbrush. Taller plants are Ocotillo and some Acacia. The view south is of Laguna Diablo and Valle San Felipe.

166   There is a beautiful cholla garden in this area. The granodiorite hills are now next to the road. They exhibit a prominent fracture pattern, are spheroidally weathered, and are cut by a light-colored swarm of dikes.

163.5   The hill to the left, Cerro Coyote, is capped by a patch of basalt.

162   There is a good view of Valle Santa Clara here. Notice how very small the alluvial fans are on the face of the high range. Their small size, for such an elevated range, indicates very recent, very active faulting. Some of the fans to the south bear slight benches that were produced by faults cutting across the fans. At about 10 o'clock, there are a series of dissected fans that have very obvious fault scarps on them.

On the floor of the playa lake there is a forest of Smoke Trees. Cheesebush is growing under the Smoke Trees. Cholla, Ocotillo, Creosote, Acacia with Mistletoe, Agave, and Desert Mallow are also seen in this area. You may even see a coyote or two.

155   Lots of Palo Verde, Smoke Trees, Cholla, Bursage, and Ocotillo in this area.

152   As the highway enters San Matias Pass, there is a gray granodioritic hill cut by numerous dikes on the right. The brown hill in front is gneiss. The highway then enters Cañon San Matias, proper, which is a very interesting low-level pass through the range considering the impressive scarps on both sides. There is very little climbing except what is needed to get up to the level of Valle Trinidad on the other side of the mountains.

DIKES CUTTING GRANODIORITE ALONG ROAD

150.5   The road cuts on the north side of the highway show the relationship between the metamorphic rocks, tonalites, and the dikes.

147   The dominant vegetation is Creosote with Ocotillo and Cholla. North facing slopes have tall Ocotillo, Agave, Acacia, Creosote, Brittlebush, and Datilillo. South facing slopes have Brittlebush, small Ocotillo, and Creosote.

San Matias Pass - This pass seems to be at the end of the Agua Blanca

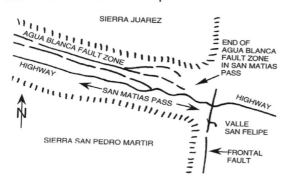

Fault Zone that was crossed near Ensenada. This fault zone is rather wide with quite a bit of movement on it and the pass is a logical place for it to go through. It now appears as though the fault simply ends here or continues through the pass as a subcrustal feature. This is unusual since it is a very strong feature with tens of kilometers of latteral offset near Ensenada.

146   The old road through this pass is in the wash on the right. It was a high-speed road in this area. There are a lot of Barrel Cactus in the pass.

143   At Valle San Matias, the Agua Blanca Fault Zone forms the north side of the pass. The straight-faced tonalite hill that comes down in a smooth slope is the fault scarp of the Agua Blanca Fault Zone. Here it shows a vertical component of the normal movement.

The vegetation consists of Datilillo, Pencil Cholla, Yucca, Creosote, Acacia, and some Mistletoe. Honey Mesquite forms a dense almost forest-like area. Keep an eye out as there are Northern Harrier hawks in this area as well as flocks of pigeons.

The **Northern Harrier** has a distinctive white rump and owl-like facial disk. Their slim bodies are grayish above and mostly white below, and have they have black wing tips. Northern Harriers inhabit wetlands and open fields. Perching low and

flying close to the ground, they search for mice, rats, and frogs.

137.4   Road to Mike's Sky Ranch (31 Km.) and the Sierra San Pedro Mártir. This road connects to the main highway Mexico 1 at San Telmo (2:140.9).

130   To the south, there is a swarm of dikes cutting the dark gray gneissic rocks. Ahead is the main area of Valle Trinidad. The hill to the north is basalt covering rhyolites and fluvial sedimentary rocks. The lower parts of the hills are lahars and fluvial sedimentary rocks. The upper part is basalt. Much of the geology is not obvious due to vegetational cover.

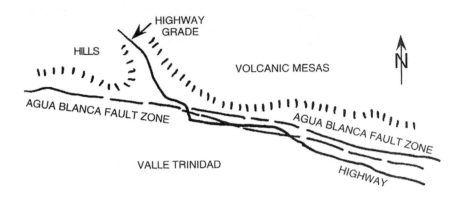

127   The highway follows directly along what is mapped as one of the traces of the Agua Blanca Fault Zone. A second trace follows along the base of the hills to the north. As the road bends, it comes directly onto the line of the fault, then bends away

124.5   Barrel Cactus, all leaning to the southeast, cover the hillsides to the right. The hills are covered with Beavertail Cactus which is a sign of overgrazed disturbed soils. There is Jumping Cholla, Creosote, and Yucca growing among the Beavertail. Creosote grows thickly in the washes.

121   The turnoff to Valle Trinidad is on one of the traces of the Agua Blanca Fault Zone. The hills to the west are composed of gneisses and tonalite. The road cuts up a small valley to climb from Valle Trinidad to cross the fault scarp onto the El Rodeo erosion surface.

119   The old road up this grade was extremely rough. Early Baja travelers who went down it were often unable to return, and had to travel through San Felipe. This is an area of mixed rock with dikes cutting through tonalite. There are zenoliths in the tonalite and numerous faults and joints. Climbing the hill, the Ocotillo is replaced by the Chaparral of the California Phytogeographic Region consisting of Manzanita, Toyon, Chemise, Oak trees, and Rabbitbrush. *(See 1:27.3).*

OLD ROAD INTO VALLE TRINIDAD

114.5   Top of the grade. The view opens to the east of the high rhyolite capped mesas.

On top of the erosion surface, there is a change in the vegetation: Junipers and Purple Sage are now dominant. This area looks like a Pinyon-Juniper pygmy coniferous forest without the Pinyon Pines. There are also Acacia and narrow-leafed Yuccas (*Yucca angosteda*). Beavertail Cactus is on the disturbed soil along the road.

108   The road leaves a small hilly area and passes onto the El Rodeo erosion surface in an area of gneisses with some tonalite. This impressive flat erosion surface was cut during the Eocene.

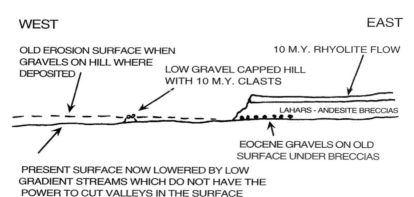

WEST

EAST

OLD EROSION SURFACE WHEN
GRAVELS ON HILL WHERE
DEPOSITED

10 M.Y. RHYOLITE FLOW

LOW GRAVEL CAPPED HILL
WITH 10 M.Y. CLASTS

LAHARS - ANDESITE BRECCIAS

EOCENE GRAVELS ON OLD
SURFACE UNDER BRECCIAS

PRESENT SURFACE NOW LOWERED BY LOW
GRADIENT STREAMS WHICH DO NOT HAVE THE
POWER TO CUT VALLEYS IN THE SURFACE

227

EL RODEO EROSION SURFACE
FROM LOW GRAVEL HILL AT KILOMETER 110

To the west at about 1 o'clock, the two low hills are capped by gravels that filled part of a stream course flowing on the erosion surface during the cutting of the erosion surface. Now due to inverted relief, they exist as hills on the erosion surface. Since these conglomerates contain some of the 10 million years old rhyolites, the streams must have been carrying the gravels sometime less than 10 million years ago. This gives a minimum rate of erosion of 100 ft./million years, or 1 mm/300 years, for this surface.

The view to the north is of the flat El Rodeo erosion surface gradually climbing into the high peaks of the Sierra de Juárez. The mountains do not look very high from this vantage point, yet they are quite steep and rugged in some places. The erosion surface was developed when this area was closer to sea level. It has been elevated and tilted to the west *(See 13:110)*.

102   Ejido Reforma. There are Meadowlarks, Ravens, and House Finches in this area. The highway is on the flat erosion surface. To the west, the hills that seem to rise from the edge of this surface are the coastal range of metamorphic mountains. To the east are the mesas of rhyolite and basalt capped lahars.

98   Scattered Pinyon Pines now join the Junipers in the pygmy coniferous forest. This area used to be covered with Pinyon Pines and in some areas Ponderosa Pines. However, the gold mines of El Alamo consumed much of this forest for the charcoal to smelt the gold and timber to shore the mines.

96   The low range of hills directly ahead represent a horst between a pair of faults that are at a right angle to the Agua Blanca Fault and seem to offset one of the branches of the Agua Blanca Fault Zone. The pond is a sag pond developed as a result of the impounded drainage along the horst.

91   Independencia.

90   The road is running approximately along a normal fault scarp, with schist on the right and tonalite on the left.

88   The road drops onto the flat, alluviated valley of Llano Colorado. Below the main hill in the far distance is the gold mining town of El Alamo. The main hill to the right of that hill is gabbro.

85.4   This turnoff leads to El Alamo (Cottonwood).

83   The road climbs into an area of granitic and metamorphic rocks. The rocks in the road cuts are sheared, altered, and cut by numerous dikes. These dikes very typically contain sphene.

As the highway leaves the Llano Colorado, the area is vegetated by the sparce pygmy coniferous forest. Rabbitbrush, Yucca, Agave, Pinyon Pines, Juniper, Laurel Sumac, Toyon, and Manzanita are typical species.

81   To the right is a reasonably straight line of low hills. These hills mark the trace of the San Miguel Fault Zone. In 1957, this fault zone was the location of a series of earthquakes. The road is still passing through schist.

76.5   The schistose nature of some of these rocks is very evident. The road leaves the flat surface and begins to descend a valley through tonalite.

74.5   Pino Solo area - In the old days, there was one large Ponderosa Pine left in this area (Pino Solo). It *was* a giant survivor of what was here before the extensive logging claimed most of the trees. Quail, Mockingbirds, and Scrub Jays are often seen along the road from Pino Solo to here.

73.5   The spring to the right along the road is at the end of a fault line. The Broom Baccharis once again become the dominant species.

71.5   The small conical peak of Cerro Colorado comes into view to the left (1/2 kilometer). It is tonalite that has weathered to a reddish color rather than the usual gray.

68   The highway crests a small rise and descends into a little valley.

Ahead is the flat crest of the Sierra de Juárez. The fault that fronts the Sierra de Juárez into Ojos Negros Valley forms the jagged slopes directly ahead. The vegetation consists of stands of Broom Baccharis, Chemise, Purple Sage Brush, Willows, and Manzanita.

67   Occasionally, there is a glimpse in the wash on the left of the old road. In the old days, this was a fine stretch of road because it was a long, sandy and smooth wash that allowed traveling at a pretty good speed. The road was lined with Broom Baccharis blocking the view of the road ahead; however, sound carried well. You could be traveling down the road at 30-40 mph and meet someone who would be traveling up the road at a similar speed. At the last moment, both of you would end up digging one wheel up out of a rut, touching a little brush, and passing each other with ease and a small wave of the hand.

59   There are ranchos along this stretch of highway that have cottonwoods and sycamores for shade and windbreaks. A running stream bordered by large willows crosses the road. This stream is from a spring that is along a fault. It's a nice rest stop. There may be some superstition however surrounding this area. Years ago as we tried to camp, some of the locals coming through spoke of spirits and ghosts and "bad blood." Sangre de Cristo is a little village near here and we weren't sure if they just didn't want us to camp or whether they were really afraid. The map indicated that there was a graveyard in this area which may be the reason they spoke of ghosts.

54   Road to Laguna Hanson.

52   The view opens to the coastal range called Cerro el Encino Solo (hill of the single oak), of meta-volcanic and granitic foothills west of a fault, and the very broad, flat, agricultural valley of Valle Ojos Negros. This valley is heavily alluviated and has extensive groundwater resources. It is bounded on the west side by a series of fault scarps and on the east side by an obscure fault, and has little topographic expression.

51   There is a riparian stream-side community along the road. There are Cottonwoods with Mistletoe in them. The natural vegetation is mostly coastal scrub with Cat's Claw Acacia, Purple Sage Brush, and lots of Mesquite.

48   The road passes into Valle Ojos Negros valley. A prominent granitic dike cuts across the road at the edge of a tonalite body. The hills to the west are variously composed of tonalite, schist, and gneiss with some granodiorite forming very knobby outcrops. Ahead on the left is the long linear scarp of one of the faults that forms the west side of the valley. The knobby appearing hill to the left with all the boulders on it is a pod of granodiorite. Most of the rest of the hills are pre-batholithic gneiss (some schist) grading into what we Ahead is the flat crest of the Sierra de Juárez. The fault that fronts the Sierra

de Juárez into Ojos Negros Valley forms the jagged slopes directly ahead. The vegetation consists of stands of Broom Baccharis, Chemise, Purple Sage Brush, Willows, and Manzanita.

OJOS NEGROS VALLEY

40   The highway bends to the left and passes by some low gneissic hills. In the agricultural fields in this area, the mesquite look more like trees than shrubs. This is because the cattle love to eat the green shoots, exposing the trunk. There is also Mistletoe in the Mesquite.

39.5   The road to Ojos Negros is to the right. The road to Ensenada continues straight to cut across a corner of the old highway. The hill ahead in the little gap is granodiorite as are the other hills to the left of the road. The hills to the right, are gneiss and plutonic rocks cut by white dikes.

37   The highway crosses the frontal fault of the range near where a diagonal fault offsets it. The erosion along the diagonal fault line probably accounts for the low pass that the road utilizes.

35.5   The hills to the right are gneisses cut by light colored granitic dikes. The granodiorite is still on the left and comes to the road at about Km 35.5

The hills are covered with typical Chaparral vegetation consisting of Buckwheat, Purple Sage Brush, and occasional Broom Baccharis with California Bays and Oaks in the stream bottoms.

34   Sharp curve. The road cuts are on the contact between the

granodiorite on one side and the gneisses on the other side. The road descends a grade through a series of mixed rocks that appear to be mostly tonalite cut by prominent, light-colored granitic dikes.

29   The road climbs through pre-batholithic metavolcanic rocks and crosses a fault. Jimson Weed, a poisonous narcotic with white bell-shaped flowers, is seen growing along the pavement's edge from here to Ensenada.

26.2   Crest of the pass and a road to the left to Agua Caliente Hot Springs (7 Km) and San Carlos Hot Springs. The main road now descends various canyons and drainages to the coast at Ensenada.

25   There is a drastic change in the vegetation. An Oak woodland grows in the narrow valleys along the side of the road. Basically, this is a riparian streamside community with Willows, Toyon, large Scrub Oak, and grasses. On the overgrazed hillsides, the vegetation is characteristic of the Coastal Sage Scrub community with Yuccas, Chemise, "Witches Hair", Yarrow, Purple Sage Brush, and Black Sage.

21   The road climbs out of Arroyo del Gallo and into a gabbro body with its typical subdued topography and limited outcrops.

19   The highway rejoins the old road. This area of tonalite is known as "Piedras Gordas" or "fat rocks," mainly because of the large, rounded tonalite and locally granodiorite boulders. For the next several kilometers, the road will pass through an area of mixed rocks consisting of gneisses and tonalites. Generally, the tonalites have the bouldery outcrops and the granodiorites do not. Close to the coast, the tonalites often weather to deep granitic soils because of the effects of moisture in the coastal weathering.

13   The road goes through a small pass for a view of the ocean, Bahía Todos Santos, Ensenada, and Islas de Todos Santos. As the road drops toward the coast, it passes through an area of tonalite that has been deeply weathered and is now covered by a rich soil.

11   The high hills to the far right are metavolcanic rocks, as are the hills off to the left. The area near the road is a bowl in the less resistant tonalites.

3   Enter the main part of Ensenada. For the best route to the Malacon, we usually take the fourth (or fifth) major street to the right after the road gently bends in a southerly direction. Both streets reach the Malacon after they cross Highway 1 (Blvd. Reforma).

Log 13 - Tijuana to Mexicali [179 kms = 111 miles]

Tijuana to Tecate - *The highway heads east from Tijuana on the elevated terraces of the Tijuana River, with higher mesas of Tertiary marine sedimentary rocks north of the river and faulted rolling hills of Eocene sedimentary rocks to the south. The steep, andesitic volcanic plug of Cerro Colorado dominates the north mesa. The highway climbs a metavolcanic hill to cross Presa Rodriguez at a narrows in the metavolcanic rocks. It then follows the marine and fluvial terraces on the south side of a gentle valley bounded by high rugged metavolcanic hills. The north side of the valley is a rolling terrace underlain by marine and fluvial sedimentary rocks and dominated by steep, conical, Pliocene andesite plugs. The highway then crosses a belt of rugged, metavolcanic hills and continues through bouldery granodiorite hills with numerous gabbro pods. It crosses over a railroad and through an area of steep tonalite hills, to reach Tecate.*

Tecate to La Hechicera - *East of Tecate, the highway passes through steep hills of gabbro, follows a steep sided canyon between tonalite and granodiorite, and then passes through the granodiorite to emerge on a flat elevated erosion surface developed on tonalite. The highway travels for many kilometers across this surface with views to the north of the Laguna Mountains and to the south of flat conglomerate capped hills.*

La Hechicera to La Rumorosa - *The highway enters a relatively flat area of rolling gneiss and schist hills locally cut by pegmatites. Between El Condor and La Rumorosa, conglomerates capping hills are exposed on both sides of an alluviated valley. Isolated hills of marble in the rolling hills of granitic rocks are being mined on both sides of the highway near La Rumorosa.*

La Rumorosa to Mexicali - *The highway begins to descend the steep gulf escarpment and drop through bouldery outcrops of granodiorite and tonalite, then through gneiss and schist cut by dikes. At the base of the escarpment, the road crosses the frontal fault and descends the alluvial fan to reach the rugged tonalite ridge of the Pinto Mountains. It then crosses a fault in a wash and climbs onto the uplifted older fan surface to continue its gentle descent on the fans toward Laguna Salada in view to the south. The highway crosses the active Laguna Salada Fault Zone at a low pass in the northern end of the Sierra de los Cucapas, a rugged complex area of granitic, metamorphic, and Tertiary sedimentary rocks. Signal Mountain is the high, rugged tonalite peak to the north. Finally, the highway descends gentle fans to the Colorado River Delta area of the Mexicali Valley.*

THERE ARE SEVERAL ROUTES FROM TIJUANA TO MEXICALLI.

You can choose to enter at the south end of Interstate 5 through El Chaparral (San Ysidro) crossing, or 8 miles east at the Otay crossing. You can also choose to take the Toll Road to Mexicali which is easily reached from the Otay crossing, or the free road which is reached from either crossing.

This log describes the multiple routes that can be taken from Tijuana to Mexicali. The shortest, least described route, and fastest route, is via the Toll Road. The most interesting, and most described route, is the route along the free road. It is recommended that you read both. The four routes described are 1) El Chaparral (San Ysidro) crossing to free road route (p.236), 2) Otay Border Crossing to free road route (p.237), 3) Free road route (p.238), 4) Toll Road Route from Otay (p.253).

EL CHAPARRAL (SAN YSIDRO) CROSSING TO FREE ROAD ROUTE

The log starts at the El Chaparral (San Ysidro) crossing. Go straight without turns to cross over the Tijuana River and be on Boulevard Rapida. Follow the south bank of the river bypassing the developed part of Tijuana. The road finally crosses the river and climbs into the metavolcanic hills. Turn right at the interchange following the Tecate signs. The road continues straight to merge with the Otay-Tecate log at Kilometer 153.

This part of the log is calculated in miles as there are essentially no kilometer markings along this stretch of road.

Mile 0   From the border, continue straight along the divided avenue heading eastward (Mexico 2) to follow the Tijuana River drainage with Pleistocene terraces on both sides. The terraces on the right (southwest) have been disrupted by faulting. The high terraces on the left are part of the mesas that stretch from the San Diego area into the eastern Tijuana region. They are Pleistocene material underlain by Miocene and Pliocene sedimentary rock.

4.4 - Home Depot old Tecate road. One of the turnoffs leads to the Agua Caliente Racetrack built on the site of the Agua Caliente Hot Springs on the active Agua Caliente Fault Zone.

Hot Springs: A hot springs is at least $10°$ C warmer than the mean air temperature. Hot springs do not normally require abnormal heat sources. The earth's temperature increases about $1.8°$ Centigrade per 100 meters of depth. Subsurface ground waters are heated to the temperature of the surrounding rock so the $10°$ warmer water need only come from about 550 meters

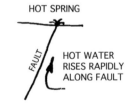

below the surface. Faults provide an avenue for the rapid rise of hot ground water to the surface resulting in hot springs. Most of the 1,000 hot springs in the western United States are along faults. Some such as Yellowstone and Lassen are near volcanic centers and are due to near surface magma.

9.3   The highway makes a bend and begins the climb up onto the metavolcanic hills. The red double-peaked mountain to the left is Cerro Colorado, a Pliocene andesite plug. This is one of a number of such plugs that stretch east toward Tecate. To the left of Cerro Colorado is the lower Cerro San Isidro, another Pliocene plug. The plug closest to the U.S. border is Cerro la Avena. Otay Mountain, which is in the U.S., is the northernmost peak.

VOLCANIC PLUGS are cylindrical masses of rock sealing the vents and conduits of volcanoes that are exposed as the more erodible surrounding rock is removed.

The hills to the right, in the sedimentary rocks of the La Mesa region, were deposited in the delta of a large Eocene river. The high hills to the left beyond the mesa are metavolcanic.

8.5   View down canyon at Presa Rodriguez, a concrete dam 1,935 feet long that was built in a narrow gorge between metavolcanic hills. The metavolcanics are exposed along the road for the next several miles.

The lake behind Presa Rodriguez is a reservoir supplying water for Tijuana. This reservoir is built along a possible fault line that is probably responsible for the alignment of the Tijuana River drainage. The low hills on the west bank of the reservoir are part of the Eocene sedimentary rocks. Ahead to the right, the higher hills are metavolcanic rocks.

10.2   Cerro las Abejas is the sharp triangular volcanic plug on the left. The white flat region west of Cerro las Abejas is called Lomas Blancas. The two andesite plugs visible to the left are Cerro Colorado and Cerro las Abejas. The highway follows a poorly defined valley in terrace material with metavolcanic rocks on the right and a terrace developed on Eocene(?) and Pliocene sedimentary rocks to the left.

10.4   **APPROXIMATE LOCATION WHERE THE TWO ROUTES JOIN** to form free road route.

If you want to follow the Toll Road to Tecate, take the Tecate Cuota turnoff and follow Boulevard 2000 about 7 miles to the Toll booth.

OTAY BORDER CROSSING TO FREE ROAD ROUTE

This part of the log is calculated in miles as there are essentially no kilometer markings along this stretch of road.

Miles

0   Border Crossing - Continue south over an overpass and through a stop sign to a signal at Industrial. (Follow the signs.)

0.9   Turn left on Industrial (a divided street in an urban area). After several miles, the highway passes along the edge of metavolcanic hills that are frequently covered by boulder conglomerates of the Redondo Formation.

4   Red Mountain and a small volcanic plug are in the middle distance to the right.

5.7   Once you cross the river, you have two choices. The branch to the left leads to the Toll Road to Tecate. It is a faster route with most of the same views. The branch to the right, and this log, turns south.

11.7   Cross another tributary of the Tijuana River.

12.5   Signed turnoff to Tecate.

APPROXIMATE LOCATION WHERE THE TWO ROUTES JOIN *(route is now in roadside Kms.)*

DESCRIPTION OF FREE ROAD ROUTE TO LA RUMOROSA

153   The conical peak (Cerro La Posta) in front on the right is another andesite plug with Tertiary terraces developed on its sides. Pre-batholithic metavolcanic rocks are now exposed on both sides of the road. The high hills to the right are metavolcanic and metasedimentary rocks. Low chaparral species are dominated by Sagebrush, Wild Buckwheat, Black Sage, White Sage, and various low annual herbage.

143   The highway enters an area of mixed granitic rocks and darker diorite cut by dikes with masses of bouldery tonalite on the hills.

140.6   The highway crosses over the railroad tracks of the Tijuana and Tecate Railroad near Los Lauelas. The railroad was called the San Diego and Arizona Eastern Railroad. The San Diego part of the line is now the San Diego Trolley red line.

139   The highway now passes through granodiorite into tonalite. The clay pits of the tile and brick operations of Tecate have been developed in the gabbro soils of this area.

136.5   Flat-bottomed native Oaks grow in this area. Cattle love to browse the oak foliage and so the trees are "pruned" as high as their necks can reach, resulting in "flat-bottomed" trees. Trees growing where cattle do not browse are shaggy bottomed with foliage reaching to the ground.

136   The Rancho La Puerta is a resort and spa.

The boulders of this area are composed of tonalite. The high hill to the left with road cuts in it is Tecate Peak on the U.S. side of the border.

133   **TECATE:** The town of Tecate is located in a bowl-shaped valley surrounded by tonalite boulder covered hills. Tecate was established in the 19th century as an agricultural center because of its abundant water and fertile soil. It is home to both the Carta Blanca and Tecate beer breweries, The Zócalo (village square) is shaded by Palm and Box Elder Trees. Follow your nose and any signs straight through town.

131.5   Just east of downtown Tecate is the intersection of Mexico 2 (Avenida Juárez) and Mexico 3 (Calle Ortiz Rubio). Mexico 3 (Log 12) heads south-westward 107 kilometers and connects with Mexico 1 on the Pacific coast at El Sauzal, 1.5 kilometers southeast of San Miguel. Highway 2 continues eastward and crosses over the tracks east of Tecate. The highway then follows the railroad eastward up a canyon.

130   Cerro la Panocha is the small round, gabbro hill that looks like the raw brown sugar (panocha) cones produced as the syrup of sugar cane is poured into cone-shaped forms. Cerro la Panocha straddles the border - its north slope is in the U.S. and its south slope is in Baja.

126   The railroad tracks can be seen across the valley to the left running along the hills. Occasionally, they are hidden from view by the vegetation and by the road cuts. The tracks turn left and head north across the U.S. border.

118.5   The granodiorite cliff ahead is Cerro Rosa de Castilla. Can you see a resemblance to Half Dome in Yosemite Valley?

CERRO ROSA DE CASTILLA

OAK WOODLANDS The three most common oaks of Baja Norte are Coast Live Oak (*Quercus agrifolia*), Scrub Oak (*Q. dumosa*), and Canyon Oak (*Q. chrysolepis*). These three oaks can be identified by studying the following chart.

Characteristic	Quercus agrifolia	Quercus dumosa	Quercus chrysolepis
evergreen	yes	yes	yes
height	30-75 feet	15 feet	20-60 feet
growth habitat	large broad crowned trees	shrub forming dense thickets	large tree forming oak woodlands below 2,000'
usual habitat	below 3.000' coastal valleys Chaparral in open grasslands	Chaparral communities	canyons, north facing slopes and in Chaparral
leaf	shiny, dark green	shiny gray-green	oblong or ovid, bluish-green
morphology	oval upper surface; hairy at vein on under sides of leaf leaf margin toothed up to 2.5' long	curled leaf, upper surface covered with brownish hairs; leaf margin toothed; ½-1' long	green above; covered with yellow or silver powder below; leaf margin toothed 1-2.5' long
acorn	slender, pointed 1-1.5' long		long-ovoid with turban shaped cup 1-2.5' long
bark	gray with broad checked ridges		smooth to scaly gray

Rancho Rosa de Castilla is located in a spectacular beautiful oak woodland area. For the next 6 kilometers, the highway passes in and out of several oak woodland valleys.

117   The hills are covered with bouldery outcrops and true Chaparral. The highway passes through broad flat valleys in the granodiorite.

111   The highway climbs a grade. For several kilometers, the highway passes through metamorphic rocks of the Julian Schist that are also cut by quartz dikes.

110 As the highway reaches the top of a long grade it begins to cross a broad flat erosion surface developed on tonalite (La Posta quartz diorite). The dominant plant species is **BROOM BACCHARIS** growing in almost pure stands. It resembles a broom.

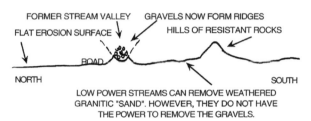

FORMER STREAM VALLEY   GRAVELS NOW FORM RIDGES
FLAT EROSION SURFACE ↓   HILLS OF RESISTANT ROCKS
NORTH   SOUTH
LOW POWER STREAMS CAN REMOVE WEATHERED GRANITIC "SAND". HOWEVER, THEY DO NOT HAVE THE POWER TO REMOVE THE GRAVELS.

EROSION SURFACES: This surface was developed during the late Cretaceous and Early Tertiary ages when this area was closer to base level. It is the effect of prolonged weathering in a more humid climate. The flat even-topped hills on the skyline to the southeast are capped with Eocene conglomerates. These conglomerates were deposited by a large river that flowed across this area during the Eocene. The source of this river was near Nogales in South-Central Arizona. It flowed across the area of the Gulf of California which did not exist at that time. The delta of this Eocene river is in the Tijuana area (*See* 15:40.5).

When streams are near their base level they cannot erode downward. Instead, they cut laterally and tend to form a flat plane. If this surface is not uplifted it will remain a flat surface. Even when uplifted, the streams on the surface do not have any power until they reach the edge and begin downcutting. Over millions of years, weathering of the surface rocks and the carrying away of the finer material by the low-powered streams lowered this surface about 1 mm per 100 years. The conglomerates were not easily removed and became resistant ridges above the surface.

239

GRAVEL RIDGE ON EROSION SURFACE

An interesting point is that once an area is left above the surface, it tends to remain drier and not weather as rapidly as the rocks on the flats. This slows the rate of erosion on the hill. As a result, the hills tend to become more pronounced and higher above the surface with time.

105.5   This turnoff leads southward to El Compadre and Valle Ojos Negros (black eyes) and meets Mexico 3 at 12:55.3.

The blue pipe brings water for municipal, industrial, and agricultural uses to this region of Baja from the Colorado River.

Red-stemmed Buckwheats are growing in the disturbed soils along the highway shoulders. The seeds of Buckwheats were ground into a meal and baked into cakes or made into a mush by native Baja Indians.

101   There is a large bend in the highway as it turns to the south and passes through the small village of Hechicera.

The Laguna erosion surface can be seen from here. Their flat concordant summits are part of the same erosion surface at about 3,600 feet higher elevation. By a roundabout route, the Laguna erosion surface descends gradually to the level of the surface near the road.

93   The stringers of white boulders exposed in the hills for the next 6 kilometers are pegmatite dikes.

89   This is a good turnout to view the vegetation of this region such as Broom Baccharis with Laurel Sumac, and Black and White Sage. Scrub Jays may also be seen in the area.

83.5   Julian schist is exposed in this roadcut.

82.8   Turnoff leads to the settlement of El Condor located on the general level of the high erosion surface. The road south from El Condor connects with the main La Rumorosa-Laguna Hansen road after about 15 kilometers.

The hills ahead to the left and to the right in the middle distance are covered with Eocene conglomerates discussed at Kilometer 109.

76.5   The hills on both sides of the road are composed of conglomerates laid down in an Eocene river. A hundred feet east of the 76 Kilometer mark there is a dirt road to the left that can be followed for 650 feet northward to exposures of the Eocene conglomerates. These conglomerates are poor in the usually present exotic meta-rhyolites due to the mixing with materials of another stream that contained no meta-rhyolite clasts. The highway is sitting on a schist and gneiss basement at this point.

The highway enters an area of impoverished Pinyon-Juniper coniferous forest. Scrub Oak, Broom Baccharis, Chemise, Yucca, Black Sage, and White Sage also grow in association with this coniferous forest. Ravens may be seen in this area.

PINYON-JUNIPER CONIFEROUS FOREST with Pinyon, Juniper, Scrub Oak, Yucca. Black and White Sage

Juniper

Pinyon Pine

Scrub Oak

Yucca

THE PYGMY-CONIFEROUS FOREST: In Northern Baja, the California Juniper (*Juniperus californica*) is associated with the Pinyon Pine (*Pinus quadrifolia* and/or *P. monophylla*). This association forms the Pinyon-Juniper community, which, due to the short trunks, is commonly called the "pygmy" coniferous forest.

The leaves of **JUNIPERS** have been reduced to bright green, imbricated (overlapping like tiles of a roof) scales that closely clothe the smaller branches. They are coated with a waxy cuticle that helps keep moisture in. Another adaptation reducing water loss are the small, dry, berry-like fruits with their protective waxy bloom. The shaggy ash-gray bark provides insulation from the desert heat, keeping tissue temperatures down, another water-conserving adaptation.

The **PINYON PINE** grows taller than the Juniper. The leaf stoma of the Pinyon are found in the bottom of pits in the needles, reducing water loss. The Pinyon has a fungus growing on its host's roots (a mycorrhizal mutualistic association). The fungal symbiont provides water for the tree and protects the roots from disease; the Pinyon provides sugar for the non-photosynthetic fungus that cannot produce its own.

Both Pinyon Pines and Junipers have been and are still useful to man and animals in Baja. People still collect the protein rich seeds, or Pinyon nuts for food. Many of Baja's birds, rodents, and deer also rely on the Pinyon nuts for food. The nuts are produced on the sporophylls (seed leaves) of the large female pine cones. Male cones are very small, producing only pollen. Today, the wood of the Juniper is primarily used to make fence posts in Baja. The cones of *Juniperus communis* are soaked in alcohol to leach the oils from the seeds and cone tissues that are used to give gin its characteristic flavor.

PINYON JUNIPER WOODLAND VOLCANIC CAPPED PEAK ON THE EROSION SURFACE

Both the Juniper and Pinyon Pine are especially plentiful on the dry slopes of the Sierra de Juárez and Sierra San Pedro Mártir below 5,000 feet in the Upper Sonoran Life Zone. Overgrazing in northern Baja is allowing them to increase in abundance and extend their domain into places formerly occupied by grasses. While reducing range lands, their increase is helping native hoofed-browsers like Deer which feed on the twigs and follage of both conifers. In addition to their wildlife food value, Junipers and Pinyon Pines also provide important protective and nesting cover for numerous bird species in Baja like Robins, Sparrows, Mockingbirds, and Warblers.

71   Marble quarried from the roadside quarry on the hill to the left and others in the area is used as decorative rock and to make cement. Several small caves decorated with stalactites have been found in the marble. This is of concern to the miners as caves reduce their marble reserves.

243

67.5   La Rumorosa. A dirt road leads southeastward from La Rumorosa, into the Sierra de Juárez, to the Parque Nacional Constitución 1857 and Laguna Hanson. The flat westward sloping erosion surface stretches to the crest of the range. The steeper gradient Gulf streams are rapidly cutting westward into the scarp. This descent of the Cuesta La Rumorosa (Cantu Grade) is the steepest and most dangerous paved road in Baja. The fault scarp that is much steeper and more evident to the south is not as steep here due to exposures of the more easily eroded schists.

The Pinyon-Juniper forest in this region is made quite scenic by bouldery granodiorite outcrops. On the slopes of the eastern drainage of the Sierra de Juárez, the granodiorite boulders are mixed with Pinyon Pines and Junipers.

64 Toll Station - The old road has been converted to a downhill toll road. Upon leaving La Rumorosa, the highway begins its steep, winding drop down the Cuesta La Rumorosa into the desert along the eastern drainage of the Sierra de Juárez. The rocks exposed at the top of the grade are granodiorite. Farther down the grade, tonalites and finally schists are exposed.

The highway passes through a small canyon of spheroidally weathered granodiorite with Pinyons and Junipers.

CUESTA LA RUMOROSA (CANTU GRADE)

RIFT VALLEYS AND THE GULF OF CALIFORNIA: The East Pacific Rise is passing under the continent and forming the Gulf of California and the Great Basin of the southwestern U.S. and northern Mexico (*See* introductory geology). This results in a series of mountains (horsts) and valleys (grabens) bounded by faults. As these mountains are eroded, broad alluvial aprons (bahadas) stretch out from the hills into playa lakes. Eventually the rise will pull Baja away from the rest of the continent along the San Andreas fault. Offset on the San Andreas fault averages 2 inches per year. At this rate, San Francisco will appear on the other side of the fault in about 15 million years. 2"/year = 30 miles/million years = 475 miles in 15 million years.

62   There is a view to the left into a bouldery valley with a small stream. The desert to the northwest in California is called the Yuha Desert. In the far distance, West Mesa is visible.

61.5   This turnout on the edge of the granodiorite pluton offers a view of the San Felipe Desert. There are several other turnouts with good views. Take your pick. Keep in mind that there is no going back to the last turnout as this is a one-way road.

Directly to the left, just barely beyond the edge of the hills is the U.S. town of Ocotillo. To the right of Ocotillo, in the far distance, the Salton Sea is visible. Farther to the right, the white spot in the nearer distance is Plaster City where gypsum is refined. Just barely in view to the right is Signal Mountain or Cerro la Centinela, a high peak in the Sierra Cucapa. The hills in the far distance, northeast of Split Mountain are the Chocolate Mountains.

59.5   First view of Laguna Salada.

58.5   A dark gray to black basalt dike cuts the granodiorite in this roadcut. This could be a feeder dike for the Jacumba Volcanics.

57   A bedrock landslide is visible at Kilometer 56.8.

56.7   The buildings on the ridge to the left are part of a pumping plant for the pipeline that pumps water over the Sierra de Juárez westward to Tijuana. The green surge towers release the air that gets trapped in the water preventing large air bubbles from building up and blocking the movement of the water as it descends along the western slopes of the Sierra de Juárez.

55   As the highway nears the bottom of the grade, the view is of Laguna Salada (foreground), Cerro Colorado (the dark linear ridge in the middle-ground), and the Sierra Cucapa with Cerro la Centinela (background). Cerro Colorado is composed of granitic rocks covered with a desert varnish.

The Sierra Cucapa are a lighter reddish-brown and lie in the background on the far side of Laguna Salada. They are also covered by a patina of desert varnish, but not as dark as the surface of Cerro Colorado.

DESERT VARNISH: A weathering feature of the desert is a thin, shiny, reddish-brown to blackish coating called desert varnish that occurs on some desert rocks. This shiny stain is only one or two micrometers thick and appears to be an amorphous gel, rich in silica and alumina that takes its color from unusually high concentrations of iron and manganese.

The chemicals of the varnish originated from atmospheric dust, the stones themselves, and underlying soil and chemicals dissolved in films of moisture formed on rock surfaces by rain, fog, or morning dew. As moisture evaporates from warming rock surfaces, it leaves ions of the dissolved chemicals behind as a coating that gradually builds up over hundreds of years. The rate of varnish formation is extremely slow.

Desert varnish occurs on the upper surfaces of rocks on alluvial fans and the mosaic pebbles of desert pavement. Similar stains are widely found around seeps, especially as dripping, large, black streaks on canyon walls. The undersides of many varnished rocks are orange (iron rich) while their sides and tops are blackish-brown (manganese rich). Since iron is less soluble than manganese, iron is left behind under the rocks as the more soluble manganese is drawn upward to concentrate on the exposed upper surfaces.

53   The highway passes through metamorphic gneiss and schist of the Julian Schist with pods of granodiorite and tonalite mixed in here and there.

The hills of this area are vegetated with Yucca, Teddy Bear Cholla, Palo Verde, and Ocotillo. This area is a good place to observe vegetational differences between north and south facing slopes.

49.7   A pegmatitic granitic dike cuts the metamorphic rocks to the left.

48   To the right, the Río Agua Grande runs through the Cañon los Llanos. There is some water as evidenced by the row of vegetation following the meander. The flora of this area is typical of a Wash Woodland vegetated by Willows, Acacia, and Creosote.

45.5   Exposures of schist in outcrop.

43.5   The highway crosses the Río Agua Grande. The predominant trees growing along the wash are Smoke Trees and *Acacia* parasitized by mistletoe. Both trees are characteristic and indicators of Wash Woodlands.

WASH WOODLANDS: Because of the limitations of water supply in the area occupied by Creosote and Bursage, the water courses, which are dry most of the year, support a characteristic flora that takes advantage of the relatively abundant supply of water during rainy periods of the winter in Northern Baja. This region of a wash and its characteristic flora is a Wash Woodland. This is a separate plant community that does not belong to the surrounding arid Creosote Bush Scrub plant community. The flora of the Wash Woodland is dominated by Palo Verde, Smoke Tree, Cat's-Claw Acacia, and Mesquite.

The seeds of many Wash Woodland species are covered by a very hard (sclerophyllous) seed coat (integument) and will not germinate no matter how long the seed is soaked in water unless the seed coat is broken (scratched through) by a process called scarification. That is, it is necessary to scratch the seed coat, allowing water to enter the seed and initiate seed germination. Scarification is accomplished by the grinding action of sand and rocks in the floods occurring periodically in the washes. This also provides the germinating seedlings with water that will supply their growth requirements during the first few weeks of germination. Flash floods also serve to disperse the seeds. Like many desert perennials, seedlings of Wash Woodland trees produce only two or three leaves immediately after germination and seemingly become dormant. However, these plants are devoting their energies to developing extensive, deep root systems that will enable them to survive long after the moisture from the infrequent floods have dissipated. Plants developing deep root systems that enable them to tap underground water sources are known as phreatophytes (phreato = well; phyte = plant).

SMOKE TREE   **MESQUITE WITH MISTLETOE**

42   The highway descends on the alluvial fan surface at the base of the Sierra de Juárez.

37   The highway crosses another of the faint meanders of Río Agua Grande developed along a fault and climbs onto an elevated part of the fan surface to pass through roadcut exposures of the alluvial gravels. This fault attests to the tectonic activity that continues today in the Imperial Valley and along the northern Gulf of California. The darker, desert varnished hills to the north along the fault are the Pinto Mountains. This area has been a popular collecting area for the colorful "Pinto Mountain Rhyolite" and petrified wood that is often polished by the wind. Be careful not to stray across the border if you explore this area.

34   The prominent peak to the left ahead is Signal Mountain (Cerro la Centinela), a high granitic peak at the north end of the Sierra Cucapa. The low hills to the left consist mostly of Tertiary sedimentary rocks uplifted along the active Laguna Salada Fault Zone.

27.7   The road south goes past Laguna Salada to the palm oases of Cañada Cantu de las Palmas and Cañon Virgen de Guadalupe.

Laguna Salada is a rift basin bounded by faults. It will continue to enlarge and will become a permanent part of the Gulf of California. Laguna Salada was nearly dry in the early 1960s, receiving runoff from the local mountains.

248

It has had considerable amounts of water in it in recent years due to major storm surges in the Gulf of California, during high tides that pushed large amounts of sea water into this basin filling it to the brim. It is supper-tidal at the south end and connects with the Gulf of California during very wet periods (*See* 14:73). The Laguna Salada Fault Zone passes close to this junction.

25.5   To the north, a large dune has developed in the lee of one of the nearby hills.

The road to the south used to lead to El Centinela campground located on the beach of Laguna Salada. The trees on this beach were underwater as evidenced by their barnacle encusted trunks. Barnacles can be seen as high as 10 feet off the ground here. The water of the laguna is very salty. Salt crystals "growing" on the surface of the shore soils are formed by efflorescence as the salt-laden water evaporates. This beach is muddy.

LAGUNA SALADA

22   The highway makes a sharp bend and winds its way through a small pass that marks a small fault and a horst of metamorphic rocks. The hill in front is desert varnished. As the road turns to the right, the vertical striations on a fault plane are visible ahead at road level.

Adjacent to the highway, the vegetation consists of tall *Acacia*, smaller Smoke Trees, Creosote, Bursage, and scattered Tamarisk trees.

19 The highway descends east-sloping alluvial fans into Mexicali Valley. The trees in this region are parasitized by Desert Mistletoe.

MISTLETOE is a woody, perennial evergreen parasite that steals sugar from its host by way of a modified stem called a haustoria. The sticky seeds are disseminated on the feet and bills of birds that eat the pearly-pink mistletoe berries.

18 There is a major Junction at this Kilometer. After 40 kilometers, this Toll Road cutoff joins the Mexicali to San Felipe road at Kilometer 9.5. Traffic from Tijuana take this road to avoid the slower traffic nearer Mexicali. It is faster; however, it is not much shorter. This road generally skirts the Sierra Cucapa on an alluvial fan apron.

15 This area was formed by the Colorado River as it built a delta across the northern end of the Gulf of California. The Imperial Valley in California is a segment of the gulf that was cutoff and isolated by the Colorado River Delta. Indio, California is 11 feet below sea level. The low point of the delta between the Imperial Valley and the Gulf of California is only 35 feet above sea level. As the delta was built, the Colorado River sometimes flowed into the now isolated Imperial Valley. This created a series of lakes alternating with dry periods as the Colorado River alternately flowed into the basin and the Gulf of California. In the Imperial Valley, the high level is marked by algal limestone along the shoreline of the ancient Lake Cahuilla at Travertine Point. However, the present Salton Sea is man-made. It is the result of an error in 1906 that allowed the flow of the Colorado River into the valley for over a year.

0 The highway diverges at a "Y" fork. The right fork is the continuation of Mexico 2 that connects with Mexico 5 in another 8 Kms. The left fork leads into Mexicali. Mexicali is to the left and San Felipe is to the right on Mexico 5.

MEXICALI, the state capital of Baja California, is the second largest city in Baja. Agriculture is the backbone of Mexicali's economy. Irrigation water for the Mexicali valley comes from the Morelos Dam on the Colorado River just south of the Mexico/U.S. Border.

DESCRIPTION OF TOLL HIGHWAY ROUTE TO LA RUMOROSA

Ties to descriptions on the free road are in Italic and brackets. It is

250

recommended that you read the much more detailed descriptions and view the photos on the free road log.

0   Otay Border Crossing. Continue south over an overpass and through a stop sign to a signal at Industrial. (Follow the signs.)

0.9   Turn left on Industrial (divided street in an urban area). After several miles, the highway passes along the edge of metavolcanic hills frequently covered by boulder conglomerates of the Redondo Formation. Red Mountain and a small volcanic plug are in the middle distance to the right.

5.7   After crossing a river, turn left to the Toll Road to Tecate.

148   The first part of this log starts at the Toll Station 148 kilometers from Mexicali. The road begins to follow a canyon in the well-jointed metavolcanic foothills with Otay Mountain and the border fence on the left.

141   The small conical hill to the right front is a volcanic plug with Tertiary terraces on the south and east sides. The road begins to pass through an area of metavolcanic and granitic rocks as it passes through rolling countryside. The bouldery outcrops are the granitic rocks.

131   Toll Road crosses free road.

129   Boulder covered Tecate Mountain is on the U.S. side of the border.

127.7   The old road in an oak-studded valley is below to the left. The tracks of the Tijuana and Tecate Railroad are near the road. (*126*)

123   Tecate – Ensenada turnoff. A gabbro hill is to the left with towers and blue pipeline carrying water from Morales Dam on the Colorado River to the Tijuana area. (*131*)

118   Turnoff.

116   Gabions (wire baskets filled with rock) used as a retaining wall.

112   Good view of Cerro Rosa de Castilla. (*112*)

110   Oak woodland to the north.

107.8   Interesting granitic roadcut. There is a horizontal shear across the middle of the cut. The granitic rock above the shear is highly weathered while the same rock below the shear is very fresh. There is a sharp boundary with no transition zone.

101   View opens up to the Eocene erosion surface, Laguna Mountains to the far north. (*110*)

92.4   Bridge over free road and second Toll Booth. (*101*)

85   The Julian Schist, of Paleozoic age, cut by numerous granitiic dikes is exposed in the road cuts along the highway for the next 10 kilometers. Good examples of dikes are exposed at kilometers 83-81 and 76.

75   The highway enters the Pinyon-Juniper coniferous forest. Scrub Oak, Broom Baccharis, Chemise, Yucca, Black Sage, and White Sage also grow in association with this coniferous forest.

68   Exit for La Rumorosa & Vallecitos. The are two volcanic capped hills to the south of the road. The volcanics are part of the Jacumba Volcanics of Miocene age (19 my).

61   The free road joins the Toll Road to become a Toll Road for the descent of the La Rumorosa Grade. (See free road log for the rest of the route to Mexicalli.) Parking turnoff and service area with attended bathroom. A small tip is always appreciated

Log 14 - Mexicali to San Felipe [190 kms = 118 miles]

Mexicali to the Sierra Pintas - This segment of the highway heads south across the flat delta sands of the Colorado River, crosses the Cerro Prieto Fault Zone, and climbs onto and begins to follow the undulating alluvial fans of the Sierra de los Cucapas. The high, rugged Cucapas to the west are uplifted fault blocks of granitic and metamorphic basement uplifted during the opening of the Gulf of California. The basalt cinder cone of Cerro Prieto and its geothermal area are to the east. South of El Tare junction, the highway crosses a delta mudflat where a low pass in the metamorphic rocks denotes the start of the almost continuous rugged granitic and metamorphic Sierra Mayor. To the left is the delta region with the Río Hardy and the Colorado River channels. South of El Mayor, the highway follows the trace of the frontal fault of the Sierra Mayor. At the south end of the Sierra Mayor the highway crosses the supper-tidal flats of the upper Gulf of California. This area is the entrance for the gulf waters into Laguna Salada.

Sierra Pinta to San Felipe - At the south end of the supper-tidal flat the highway enters the highly faulted and very rugged Tertiary volcanic area of the Sierra Pinta. The low steep hills to the left of the highway at the north end of the Sierra Pinta are calcareous Paleozoic sedimentary rocks. The dune field was blown from Laguna Salada. The highway then travels through several low passes in the volcanic rocks passing remnants of uplifted and dissected alluvial fans. South of the Sierra Pinta, the highway undulates along the lower edges of uplifted alluvial fans to the junction of the Ensenada highway. The supper-tidal salt flats of the Salinas de Omotepec dominate the view to the east. Close to the junction, the high rugged hills on the right are the granitic and metamorphic rocks of the south end of the Sierra Pinta. The rounded volcanic hill on the left is El Chinero. South of El Chinero Junction, the highway continues to traverse the uplifted Alluvial Fans along the edge of the Salinas de Omotepec to San Felipe.

0   If you cross the border at Mexicali, follow the signs (route varies) to San Felipe on Mexico 5. This *highway* log starts at the Junction of Mexico 2 and 5 at Kilometer 0. This point is some distance south of the border.

SAN FELIPE DESERT: The San Felipe Desert is one of four sub-deserts of the Desert Phytogeographic Region. Generally, the sparsely vegetated, extremely arid San Felipe Desert extends south from the Border along the eastern escarpment of the Sierra de Juárez and the Sierra San Pedro Mártir to Bahía de Los Ángeles. This desert is extremely arid because it is in the rain shadow of the peninsular ranges. The two dominant plants of this region are Creosote and Bursage that comprise the Creosote Bush Scrub plant community *(See 14:108).*

2   Rio Nuevo (New River in the U.S.), a spillway gate, and a pond are on the south side of the highway. Río Nuevo flows north into the Salton Sea.

9.5   There is a major Junction at Kilometer 9.5 where the Toll Road cutoff from the Tijuana to Mexicali joins the Mexicali to San Felipe highway. Traffic from Tijuana can take this cutoff at 13:18 to avoid the traffic congestion nearer Mexicali. It's faster, however, it is not much shorter.

The mountains to the west are the Sierra Cucapa. To the far right, the higher ranges of the Sierra de Juárez are visible.

16.6   This canal marks the trace of the Cerro Prieto Fault zone. The 820 foot high dark peak to the left front at 10 o'clock is Volcan Cerro Prieto, a Quaternary volcanic cone with uplifted Tertiary sedimentary rocks on its sides. The Mexican government generates 720 MW of kilowatts of power from geothermal wells ($315^{\circ}C$) located southeast of Cerro Prieto. These geothermal power plant facilities are the second largest in North America, exceeded only by the "Geysers" geothermal facilities located north of San Francisco, California. There are hot mineral springs and mud volcanos in the area near a swampy area called Laguna Volcano.

21.2   Streams are eroding into the alluvial fan to the right, establishing a new gradient at the new base level about 10 to 15 feet below the former base level. This has created a series of sharp arroyos and abandoned fan terraces in the sediments.

24   The vegetation of the San Felipe Desert is uniformly low and sparse. The flora of this region is predominantly Acacia, Smoke Trees, Bursage, and Ocotillo. The Acacia and Smoke Trees are heavily parasitized by mistletoe (See 16:15.5). Near the Sierra Cucapa, the vegetation becomes taller. The quarry on the right is developed in granitic alluvial sands.

25   The highway approaches and skirts the Sierra Cucapa. The Sierra Cucapa were uplifted along several fault zones that bound and diagonally cut the range. In the central part is largely tonalite and granodiorite.

37   El Tave junction is the intersection of Mexico 5 and BC 18. South of this junction, the Sierra Cucapa are largely composed of granodiorite, prebatholithic schists, and calcareous metasedimentary rocks.

42   The Sierra Cucapa has declined into a low range of foothills. To the right front is a low pass that separates the Sierra Cucapa from the Sierra el Mayor which is the main mountain mass ahead.

49   At El Medanito, the highway curves southeast along the base of the Sierra el Mayor.

49.5   The metamorphic rocks of the Sierra Mayor are close to the highway. The sloughs and banks of Río Hardy provide excellent bird watching.

50   The hills on the right have a series of smooth sloping small terraces developed on them. Each level represents a different episode of uplift of the Sierra el Mayor.

55   As the highway skirts the Sierra el Mayor, it passes an abandoned backwater slough of the Río Hardy. The river has changed its course many times and may return to revive this abandoned slough at some future date.

56   The vegetation of this area is typical of slough vegetation and is dominated by flat, narrow sword-shaped monocot culms (stems). Tamarisk, Broom Baccharis, Acacia, and Palo Verde are growing on the slopes.

66   Due to the high salinity of the soils in this area only a few halophyte species are able to grow. Consequently, the vegetation to the east is dominated by almost pure stands of halophytic Creosote and Bursage, sparsely interspersed with Tamarisk.

69   The highway traverses a broad flat saltpan. The range ahead is the Sierra las Pintas. To the right of the highway, the Sierra las Tinajas, a lower range of mountains, are also visible. The higher range farther to the west of the Sierra la Tinajas range is the flat crest of the Sierra de Juárez.

73   The highway crosses the supper-tidal flats of the Gulf of California on a levee. This area is actually the southeastern end of Laguna Salada, a basin located to the northwest (See 13:27.7). The high mountains to the west are the Sierra de Juárez. To the south are the Sierra Pintas.

80   A series of active and partially stabilized sand dunes are located along both sides of the highway. Creosote and Bursage are growing on the dunes. On hot days, mirages are commonly seen in this area on the salt pans around the Sierra las Pintas.

83   The Sierra las Pintas are the higher mountains to the right ahead. They are a thick sequence of Tertiary volcanics. The hills to the left ahead are upper Paleozoic meta-sedimentary rocks that contain Crinoid fossils.

87   A dune field has developed on the foothills of the Sierra las Pintas. They originated as winds blowing southwest across Laguna Salada picked up sand and deposited it on the eastern slopes of this range.

PALEOZOIC ROCKS WITH SAND DUNES

97.3   The highway travels along the eastern edge of the Sierra las Pintas for many kilometers. The broad white flat area to the left is part of the Salinas de Omtepec, a supper-tidal flat located at the northern end of the Gulf of California. Supper-tidal flats are inundated during very high tides.

105   La Ventana.

THE CREOSOTE BUSH SCRUB PLANT COMMUNITY OF THE SAN FELIPE DESERT: The highway passes through the Creosote Bush scrub plant community of the San Felipe Desert area of Baja's Desert Phytogeographic Region. This community is found below 3,000 feet on slopes, alluvial fans, and valleys in the desert. The precipitation is low and mostly comes in the spring with a few summer showers. The pedocal soils that support this community are usually well drained, and the extremes of temperature are great. The community is so named because Creosote Bush (*Larrea divaricata*) is the largest most abundant plant. Along with Bursage (*Franseria dumosa*), these two plants compose nearly 90% of the foliage cover.

Creosote Bush is a shrub that is commonly 3-6 feet high with very small resin-covered evergreen leaves. **Bursage** is a low rounded, ashy-gray-green shrub 8 to 24 inches high with grayish-white bark and stiff intertwining branches. (See 3:122)

Creosote Bush and Bursage are the two dominant shrubs in the **Desert Phytogeographic Region**. They look as though they have been hand planted in straight lines. However, they are naturally spaced as an adaptation for utilizing water more efficiently. Every plant requires a certain amount of water. In wet soils, plants grow closer and more randomly; in dry environments, plants are spaced almost territorially. They grow evenly and are widely spaced to ensure that each plant receives its required amount of water and minerals. The drier the environment, the wider and more regular the spacing. This is easily seen by noticing how widely spaced and short the Creosote Bush is on the tops of the hills in this region. On the slopes, Creosote Bushes become more numerous, more closely spaced, and taller. Even taller and more closely spaced Creosote Bushes grow at the bottoms of the hills where runoff collects. The tallest Creosote Bushes grow along the shoulders of the highway where they receive the runoff from the highway.

Another factor responsible for the linear arrangement of Bursage and Creosote Bush is that they grow along tiny drainages, known as rills, that are less than two inches deep. Small rills in desert landscapes may alter vegetational distribution patterns and community composition.

A third phenomenon is allelopathy. If seeds of the Creosote Bush or other plant species were allowed to grow under or between the Creosote Bush, they would compete for the meager soil water and minerals available in the extremely arid nutrient-poor desert soils of the San Felipe Desert. To prevent competition, resinous leachates from the Creosote Bush's leaves and stems "poison" the soil, which prevents the growth of other plants.

Despite the barren appearance, quite a variety of desert species grow here forming a regional assemblage of plant species known as the **Creosote Bush Scrub**. They are:

CREOSOTE BUSH SCRUB	
PLANT COMMON LOCATION	
Bursage or Burroweed	Widespread-usually
(*Franseria dumosa*)	with Creosote Bush
Creosote Bush (*Larrea divaricata*)	Slopes, washes
Ocotillo (*Fouquieria splendens*)	Slopes, flats, washes
Cheesebush (*Hymenoclea salsola*)	Washes, highway edges
Cholla (*Opuntia* sp.)	Rocky slopes, disturbed soils
Desert Willow (Salix sp.)	Infrequently along watercourses

257

Other types of vegetation, that locally grows on the eastern foothills of the Sierra las Pintas consists of Brittlebush, Candelabra Cactus, Mesquite, Spurge, Cat's-Claw Acacia, Smoke Tree, Desert Thorn, and Palo Verde.

123   The Sierra Pintas are highly mineralized. The La Fortuna and Moctezuma mines are located in the hills to the right. Silver and gold are extracted from both mines. The mine dump for these mines is to the right.

134   The volcanic hill to the east is Cerro el Chinero. The flats are covered by coarse materials that have formed desert pavement.

DESERT PAVEMENT: Deflation (from the Latin meaning "to blow away") is

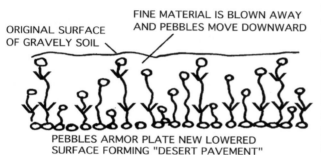

ORIGINAL SURFACE OF GRAVELY SOIL

FINE MATERIAL IS BLOWN AWAY AND PEBBLES MOVE DOWNWARD

PEBBLES ARMOR PLATE NEW LOWERED SURFACE FORMING "DESERT PAVEMENT"

an erosive process by wind that blows the sand and dust particles away and leaves the larger pebble or cobble sized particles. Aided by alternating wetting and drying of the ground, the pebbles slowly rotate, settle, and concentrate into a flat armor-like mosaic called desert pavement. In some regions of Baja, patches occur where every stone is a different color. The pebbles that form the desert pavement on the lower slopes of Cerro el Chinero are a uniform brownish color. The even surface of the mosaic floor no longer offers any elevations that the wind can attack, further deflation ceases, and the erosion process ends.

135   The high ridge to the right in the distance is the crest of Sierra San Pedro Mártir. Its high peak, Picacho del Diablo (Elev. 3,115 m., 10,219 ft.) is located in Baja's "Parque Nacional San Pedro Mártir Constitución, 1857".

Dense **CHEESEBUSH** and large Brittlebush are growing along the highway. If you crush the leaves of the Cheesebush between your fingers they produce a cheese-like odor.

The amphitheater in the center of Cerro el Chinero contains the volcanic cinders of an eruptive center. At the base of the hill, sand dunes that have formed against the hill have been dissected by streams which gives the impression of cone-shaped alluvial fans. The dunes on the flank of Cerro el Chinero are sparsely vegetated with small tufts of grass, scattered Brittlebush, and Skeleton Weed. Cerro el Chinero, itself, is covered by dense stands of Ocotillo.

140   Junction of Highway 3 to Ensenada (195 kilometers). If you are heading to Ensenada turn to the right. (See log 12).

157   First view of the Gulf of California and the Salinas de Omtepec.

161   The shoreline of the Gulf of California approaches the highway from the Colorado River Delta area. The slope that the highway is on is graded to the flat area of the salt-pans of Salinas de Omtepec.

The Salinas de Omtepec are laced with little sloughs that remove the water after tidal flooding occurs during high tides. They are supper-tidal flats that continue along the coast to the head of The Gulf of California. They are related to the 30-foot tidal range in the upper gulf and the emptying of the Río Colorado into the upper Gulf of California.

165   The brown hills to the right are desert varnished tonalite. The vegetation is similar to the flora or the Anza-Borrego Desert. The predominant plants are Ocotillo, Garambullo, abundant Mistletoe, Cat's-Claw Acacia, Mesquite, and Purple Bush. Opportunistic types of vegetation such as Cheesebush are growing along the side of the highway. It is almost the only green plant in this entire area.

167   To the east of the highway is Cerro el Moreno, a Miocene volcanic plug. Nearby there is a small inlet on the coast, Estero Primero, in the southern region of Salinas de Omtepec.

PREDATORY HAWKS OF THE DESERT: The **Red-Tailed Hawk** is a well-known buteo commonly seen perched on top of telephone poles or fence posts and where there are open hunting grounds. Usually hunting alone, this hawk may sit motionless for hours, then suddenly emit a high scream and swoop down to prey on rabbits, rodents, and other ground dwelling prey. Two field characteristics that make red-tails easily identifiable are their uniformly colored red tail and their dark belly-band.

THE AVIFAUNA OF THE SAN FELIPE DESERT: The bird species of the Gulf coast of Northeastern Baja are the same as the Californian Region, the desert regions of mainland Mexico, and the Southwest U.S. The following sixteen species are typically seen along the northeastern peninsular Gulf shoreline and in the Creosote Bush plant community of San Felipe Desert between the International Border and San Felipe:

BIRD NAME	LIKELY LOCATION
Amer. White Pelican	Fly low over coastal waters
American Kestrel	On telephone wires
Anna's Hummingbird	Near red or yellow flowers
Bendire's Thrasher	On the ground in the brush
Cactus Wren	Flying or perching on cacti
Calif. Brown Pelican	Flying over coastal waters
California quail	On the ground in coveys
Coasta's Hummingbird	Near red or yellow flowers
Gila Woodpecker	On poles or in large plants
Gilded Woodpecker	On poles or in large plants
Greater Roadrunner	crossing the highway
Ladder-back Woodpecker	On poles or in large plants
LeConte's Thrasher	On the ground in the brush
Red-Tailed Hawk	perching on poles or soaring
Turkey Vultures	On carrion or soaring overhead
Western Meadowlark	Fences bordering meadows

170   The high hill to the west in the near distance is Cerro el Colorado composed of Tertiary volcanics. The lower hills in front are all tonalite. The low darker-gray hills to the southwest are pre-batholithic slates and schists. In the distance are the high peaks of the Sierra San Pedro Mártir.

177   For the next several kilometers, the highway passes semi-stabilized dune fields of sand blown from the sandy beaches of the Gulf. There is a view of the Gulf of California, and if the weather is clear, Isla Consag, a steep sided volcanic peak may be visible in the middle of the Gulf 18 miles offshore.

187   The highway passes through a low range of hills of faulted pre-batholithic carbonate rocks. They are covered by desert varnish that gives their surface a very dark appearance (See 13:55).

189   As the highway enters San Felipe, Cerro el Machorro and the point of Punta el Machorro that are composed of tonalite and granodiorite are to the left. The rocks to the left at the monument circle are pre-batholithic.

Log 15 - Ensenada to Tecate [114 kms = 71 miles]

Note: At the time of the printing of this edition, the Ensenada to Tecate highway was undergoing a series of major realignments to promote a smoother flow of traffic. There are new road cuts and older road cuts have been eliminated. The road is now shorter and the kilometer markings have not been moved. The kilometers noted in this log are the existing kilometer markings on the side of the road.

The highway travels up a gentle alluvial valley flanked by steep hills of the marine Rosario Formation, then climbs a steep grade into the rugged metavolcanic hills. After passing through the metavolcanic hills, it travels through a rolling bouldery area of tonalite on the south side of the Valle Guadalupe, eventually travelling on the alluviated floor of the valley with steep meta-volcanic and granitic hills in the distance on both sides. It then travels through more granitic rocks and crosses a fault into an area of rugged hills in slatey metamorphic rocks with scattered granitic rocks. The highway enters steep bouldery tonalite hills near El Testerazo and climbs a grade through the conglomerates of the Las Palmas Gravels.

The highway then descends a grade through the conglomerates to the alluviated Las Palmas Valley. North of the valley, the highway climbs through steep tonalite hills and then through the bouldery granodiorite. It follows a spectacular contact between a dark, smooth, gabbro hill on the right and a light, bouldery, granodiorite hill on the left, then travels through the rugged granodiorite hills past several gabbro hills to descend steeply through tonalite into Tecate.

105   This log begins at the junction of Mexico 1 *(See 1:101.3)* and Mexico 3 and ends at Tecate *(See 13:131.5)*. The highway initially goes inland in the alluviated valley of Río San Antonio. It is bordered by hills of Cretaceous gravels of the Rosario Formation and then metavolcanic rocks on both sides.

102   Intersection with bypass road that circles around Ensenada.

101   The highway begins to climb a grade through metavolcanic rocks with frequent reddish cobble to boulder conglomerates of the Redondo Formation that represent a period of intense uplift and erosion during the Cretacous age.

REDONDO BOULDER CONGLOMERATES UNCONFORMABLY OVERLAIN BY TILTED AND FLAT BOLDERY SANDS

98.7   The roadcut at the top of the grade exposes a massive purple andesitic breccia, a common rock type in the metavolcanic rocks. The highway passes through the metavolcanic rocks for the next 2 kilometers.

96   The highway enters the southwestern end of a valley in the Oak woodland. It passes through numerous Oak woodlands intermixed with chaparral areas all the way to Tecate.

94.5   San Antonio de los Minas.

91.5   The road to the left to Alegre (El Tigre) connects with the Highway between Tijuana and Ensenada. The boulder-covered slopes are tonalite and the smoother slopes are generally metavolcanic rocks and slates.

88   The flat hill and ridge to the left are a remnant of an older alluvial fan. This ridge converges with the highway at Kilometer 87.

85   Views of the main part of Valle Guadalupe begin to be seen to the left. The view opens up about Kilometer 83. The reserected Olive grove is reputed to be one of the largest in Mexico. The hills on the far side are prebatholithic metavolcanic rocks and tonalite.

77.4   The highway crosses Río Guadalupe which originates in the Sierra de Juárez near Laguna Hansen and meanders through Valle Ojos Negros and Valle Guadalupe to the sea at La Misión. Most of the water flow occurs underground leaving the streambed surface dry. The vegetation in the streambed consists of Tamarisk, Broom Baccharis, and Willow.

GUADALUPE VALLEY AND TONALITE HILLS

75.5   The road crests a small rise with a sweeping view of Guadalupe Valley.

71.5   The highway follows a fault line valley in tonalite. Most of the road cuts in this area are highly weathered

70   For the next several kilometers, many small Riparian Woodland areas are seen along the right side of the highway. The dominant vegetation along the stream banks is Coastal Live Oaks, Scrub Oak, Sycamores, and Willows. Acacia, Broom Baccharis, Chamise, Toyon with Dodder, and Wild Buckwheat are the plants that grow on the low surrounding hills.

68 to 63   Major new road cuts of sheared and faulted pre-batholithic graywackes, and argillites cut by dikes are exposed along the highway. These outcrops resemble the Triassic(?) to Jurassic Bedford Canyon Formation of the northern Santa Ana Mountains in California. They are slightly metamorphosed, highly fractured, sheared, and faulted sediments partially derived from a metavolcanic terrane. To the east, there are outcrops of Ordovician quartzite, chert, argillite, and marble in an allochthon (Lothringer, 1984).

DIKE IN ARGILLITES

ROAD CUT WITH SHEARED ARGILLITES (GRAY) CUT BY WEATHERED GRANITIC DIKES (BROWN).

63.6   This road cut exposes the sheared slates and argillites (gray) cut by weathered granitic dikes (brown). The slates and argillites have numerous steeply dipping sheared areas.

BOTH THE SLATES AND ARGILLITES AND GRANITIC DIKES ARE LOCALLY FAULTED.

63   Entering the valley of Ignacio Zaragoza where the highway passes through a Coastal Sage Scrub community that consists of Rabbitbrush, Scrub Oak, and several species of composites. The large green trees along both sides of the highway are Tamarisk, Cat's Claw Acacia, Cottonwoods, and Willows.

The major northwest-southeast-trending Vallecitos Fault Zone crosses the highway near the south end of the valley. This fault zone may continue northward through to the Tijuana River Valley. The road cuts for the next 5 kms. expose dark-colored slatey to phyllitic prebatholithic rocks.

62   A beautiful park like oak woodland extending along the drier banks of a Riparian woodland is now visible to the left side of the highway. The large silver-barked trees are Sycamores; the dark-green trees are Oaks.

62.5   The highway passes through road cuts with good exposures of schist and gneiss cut by dikes.

60   The highway has entered the broad alluviated Vallecitos which is ringed by metavolcanic hills with a few scattered patches of tonalite. Dikes cut some of the hills.

55.4   The lower part of the small valley ahead is vegetated with riparian cottonwood trees that are heavily parasitized by mistletoe.

52   The hill of conglomerate ahead has been uplifted between two

branches of the Falla de Calabazas. The highway has been built in tonalite. Numerous small shear zones cut the rocks for the next kilometer. The road turns left to follow the fault valley.

TONALITE BOULDER

The valley at Kilometer 51.2 is a fault-line valley. The highway runs parallel to the fault zone at the Kilometer 51 mark.

50.3   The highway crosses Río Las Calabazas. This fault-line valley parallels the one at Kilometer 52.

49   El Testerazo. The hills ahead are composed of Eocene conglomerates that were deposited 45 million years ago by a major river that flowed across the range from Sonora to the ocean in the Tijuana area. The high hills to the left are metavolcanics.

47   As the highway climbs a grade, it passes through Eocene conglomerate outcrops.

There is a significant shortening (approximately 10 to 12 kilometers) of the highway between here and Kilometer 32.

As this old road reaches the flat summit, the view to the left is of the erosion surface and the top of the conglomerates of the Eocene river. The conglomerates that comprise these intervening hills are part of a string of conglomerate exposures that stretch to the crest of the Peninsular Ranges to the east. They contain a small percentage of rhyolite clasts similar to those of the Poway Conglomerate of the San Diego area (Minch, 1972, 1984). The conglomerates rest upon an Eocene or older erosion surface. The erosion surface can be seen to the east (See 13:111).

CONGLOMERATES FROM EOCENE RIVER ON SOUTH GRADE

32   The Falla de Calabazas that the road crossed near El Testerazo is seen near the break in the slope ahead. It passes through the notch between the two very high hills, and across the valley to the right of the zigzag in the Cerro Bola road. The vegetation of this area is the Coastal Sage Scrub plant community (See 1:14.8).

32   Rancho Viejo - *The kilometer marks are correct from Tecate to here. They will probably be moved from this point south. Use approximate landmarks and your odometer.*

At this point, it is worth taking a **detour** up the old road for a view of the Eocene Erosion surface. (This detour loups to the highway about 2 kms. to the south.) As you reach the top, the vista opens to the tonalite hills cut by several light-colored granitic dikes. The light-colored linear bouldery outcrops are erosional remnants of these granitic dikes. Two granodiorite peaks form the skyline: Cerro los Bateques is the lower peak and Cerro La Libertad is the higher peak.

AREA OF TONALITE OUTCROPS

EROSION SURFACE *Cerro La Libertad to the northeast was a monadnock on this surface. Monadnocks are hills representing scattered erosional remnants on a low-relief erosional surface (peneplain). West of this area the surface has been largely destroyed by stream erosion.*

EDGE OF EOCENE RIVER CHANNEL

31   For the next 3 kilometers, the highway passes along the edge of Valle de Las Palmas.

30.5   The road to the left climbs Cerro Bola where Microondas Cerro Bola is located. The lower part of the road crosses a smooth gently sloping surface that is part of the old alluvial fans that developed in this area before it was uplifted. The upper part is in the metavolcanic rocks.

29   Las Palmas.

27.5   The highway crosses Arroyo Seco on Puente Valle de Las Palmas.

22   The highway passes through an extensive area of bouldery outcrops of granodiorite.

17.5   Crest of grade and an area of mixed light-colored granitic and dark-colored metamorphic rocks.

16   Notice the contact, along the highway, between a granodiorite hill to the west and a gabbro hill to the east. The gabbro weathers easily and has no outcrops, while the granodiorite hill weathers more slowly and has bouldery outcrops. As the gabbro weathers, it forms clay soils. These clay soils protect the underlying gabbro from further weathering; this results in fresh gabbro relatively near the surface. However, the granodiorite weathers to a sandy soil that allows further deep weathering leaving the surface boulders on a weathered base.

269

GABBRO VS GRANODIORITE OUTCROPS

7   The metavolcanic peak in the distance is Tecate Peak in the U.S.

3   The descent into Tecate continues as the road passes under the new Tijuana to Mexicalli Toll Road. The numerous brickyards in this area exploit the clay soils developed on isolated gabbro bodies.

0   The highway crosses railroad tracks in Tecate on Calle Ortiz Rubio. The main Tijuana to Mexicali highway. Those who wish to continue directly to the United States should follow the signs. In 2017, you turned right on the main road for some distance to a marked junction then turned left to the border and then left along the border. There were two gates at the Tecate border crossing. This log joins the Tijuana-Mexicali log at Kilometer 131.5.

While this area seems to be arid, the world's record for high intensity rainfall of 11" in 80 minutes was set in the small U.S. town of Campo just north east of Tecate. Campo was a rail camp when they were building the San Diego & Arizona Eastern Railroad and a U.S. Cavalry training post.

Log 16 - L.A. Junction to Bahía de Los Ángeles [66 kms = 40 miles]

The highway gently climbs eastward on the flat surface of the Miocene fluvial sedimentary rocks toward a low rolling series of metamorphic hills. It then drops into an alluvial wash and climbs out to pass south of rugged rhyolite hills, then between a series of rugged metamorphic hills in a broad sloping alluvial valley. The hills and mesas to the north are capped by rhyolite and basalt. The highway then steeply descends a pass in metamorphic rocks to skirt the south side of the dry lake of Agua Amarga along a series of rugged metasedimentary hills.

The opposite sides of the playa are a steep, rugged series of faulted, metamorphic and granitic rocks capped by mesas of basalt tuffs and rhyolite. After crossing the southern part of the lake, the highway again descends a steep canyon in rugged hills of metasedimentary rocks to the alluvial fans on the shore of Bahía de Los Ángeles. Here the highway turns south on the fans and parallels a fault scarp in the granitic and metasedimentary rocks.

0   The junction of the highway and the paved road that heads east to Bahía de Los Ángeles is 10 kilometers north of Punta Prieta. For the first ten kilometers, the highway climbs on the Miocene fluvial sedimentary rocks that mantle this relatively flat area. The mesa to the right consists of volcanic and fluvial sedimentary rocks capped by dark basalt. As the highway continues eastward it approaches Mesa La Pinta, a granitic and metamorphic hill. The vegetation consists of Vizcaino Desert flora (*See 3:74.5*). Cirio, Ocotillo, Elephant Tree, and Cardon are the dominant taller plants. Datilillo, Yucca, Cholla, Garambullo, and desert annuals form the low "understory." There is an occasional Palo Verde or Mesquite heavily parasitized by Mistletoe.

7   Hills of rounded granitic rock and metamorphic rocks can be seen about a mile to the left of the highway.

10   The highway enters an area of low metamorphic and granitic hills. Many of the granitic rocks are spheroidally weathered.

14.5   The highway dips into a wash vegetated by Datilillo and Cardon.

15.5   As the highway quickly climbs back onto the surface of the Miocene fluvial sedimentary rocks, epiphytic Ball Moss covers the Cirio. Desert Aster, Atriplex, Mesquite, Desert Hollyhock, Coyote Melon, Indigo Bush, Agave, Elephant Tree, and Ocotillo also grow in this area.

19   The pinkish colored flat-lying hills of the Mesa Tinajas Coloradas to

the left are Miocene rhyolites overlying Miocene fluvial sedimentary rocks.

18   For about one half kilometer, the highway skirts a dark colored, gabbro hill to the left of the highway. (Very few gabbro hills are accessibly exposed along Baja's highways.) gabbros are dark colored basic, intrusive igneous rocks that are the intrusive equivalent of the extrusive basalts.

23   The vista opens up ahead, as the highway descends a broad alluvial plain on the Pacific coast drainage, into the Valle Agua Amarga with beautiful, thick stands of an Elephant Tree forest. Although the Elephant Trees look "dead" most of the year, during the rainy season they put out leaves and look lush and green. When they flower, they look like they are capped by a pinkish haze. The well-drained drier slopes in this area are covered with them. Cirios and Cardons vegetate the lower, flatter elevations.

30.3   As the highway crests a low pass it enters the Gulf drainage. At the crest of the pass, the vegetation consists of Elephant Tree, Datilillo, Garambullo, Agave, Greasewood, Cirios, and Cardon. The dominant tall plant here is the thick-trunked Elephant Tree. The parisitic dodder mantles some of the Elephant Trees for the next several kilometers.

32.5   The hills directly ahead are pre-batholithic metasedimentary rocks with resistant light colored layers. The fine-grained, white-bedded layers along the highway are Quaternary lacustrine sedimentary rocks that may represent dissected remnants of a tilted ancient lake bed.

33   The vista opens to Playa Agua Amarga (bitter water) which refers to the highly saline, unpotable waters found in the valley. The Sierra Toro, located on the far side of the valley, are a mixture of granitic, metamorphic, sedimentary, and volcanic rocks.

PLAYA AGUA AMARGA

PLAYA LAKE: Playas are the flat, dry area of an undrained desert basin underlain by clay, silt, and commonly soluble salts. It may be covered by a shallow, intermittent (ephemeral) lake during the wet season. When the water of the lake evaporates, it leaves a playa covered by salt deposits.

Valle Agua Amarga was historically a deeper valley, but valleys get filled up over geologic time. The deep valley fill is composed of the silt, salt, and sand that was originally weathered, eroded from the bedrock of the surrounding highlands, and deposited in the valley, a process that is still occurring.

The highway skirts the Playa Agua Amarga salt pan. The highly saline, white, central portion of the playa is totally devoid of life. However, the less saline soils along the edge of the playa are vegetated by hearty salt tolerant halophytes (salt lovers).

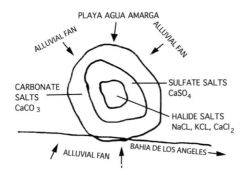

SALT PAN: Salt (CaCl$_2$, KCl, NaCl) and mud make up most of the flat floor of the Valle Agua Amarga. However, several other salts are mixed in with the Halite and mud in variable proportions. Originally rainwater dissolved all of the various types of salts from the bedrock of the surrounding mountains. Runoff and groundwater seepage have slowly moved the salts downward toward the lower parts of the valley floor resulting in the formation of the salts in the playa. Different salts move at different rates and so end up in different localities depending on their solubility in water.

The salt in the soils of the alluvial fans is predominated by the least soluble carbonate Calcite (CaCO$_3$). Calcite crystallizes from groundwater as a natural cement and is commonly visible as white seams.

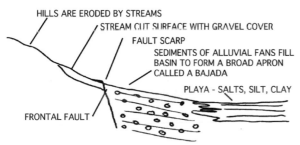

Moving down the fan, to the edge of the playa, the slightly more soluble Gypsum (hydrous CaSO$_4$) becomes abundant both as caliche-like veins and as gypsum crystals that formed by efflorescence.

273

Efflorescence is the process by which crystals grow due to the evaporation of salt-laden water. The central parts of the Playa Agua Amarga are mostly Halite (NaCl) and chlorides ($CaCl_2$, KCl) since they are the most soluble salts. Thus, there appears to be a crudely concentric zonation around the salt pan, reflecting increasing solubility, from carbonates at the top of the slopes to sulfates and finally to the highly soluble chlorides concentrated on the barren, lifeless floor of the salt pan.

35.5   The rocks along the side of the highway are pre-batholithic metasedimentary rocks and schists.

WHY IS PLAYA AGUA AMARGA NEARLY DEVOID OF VEGETATION?
Plants are extremely sensitive to changes in water quantity and quality. A sudden decrease in plant height and/or a change in the species composition of an area can give clues as to the quantity of water available and the quality of that water. Plants needing permanent water supplies, such as willows, alders, ferns, and rushes, are called phreatophytes (well plants). Plants needing little water to survive, such as Baja's Mesquite, Greasewood, Palo Verde, and various cacti, are called xerophytes (dry plants). No matter how dry the soil, there will be some plants that have adapted to the low level of water. However, few if any, plants can tolerate more than 6% soil salinity. The high salinity of playa soils is the result of runoff, carrying dissolved salts that collect and concentrate in the low, central portions of playas.

On Agua Amarga, much of the ground is devoid of vegetational cover. This indicates that the salinity of the soil is greater than 6%. There may even be standing water, but with a salinity of greater than 6% a "physiological desert" exists to which no plant can adapt. Notice that a few plants are seen growing on low hummocks in the playa and along the margins of the playa. This indicates that the soil salinity there is less than 6%.

39   Notice the dramatic vegetation change that occurs as you look from the slopes down onto the playa salt pan. These changes reflect slope drainage and increasing soil salinity. The hills to the right are covered with Elephant Trees; the lower slopes are covered with Cardon, Jumping Cholla, and Ocotillo; the flats are densely vegetated with Creosote Bush, Teddy Bear Cholla, and Greasewood. The very Saline Salt flats are bare.

41   Dense stands of Smoke Trees are growing in the wash on both sides of the highway. Smoke Trees only grow in washes since their seeds must be scarified by the sands of the wash during wet periods and water is necessary to leach out germination inhibitors *(See 13:43.5).*

42.5   The sign here indicates there is a back way to Misión San Borja

274

recommended for 4-wheel drive vehicles only. Because of better road conditions, the road from El Rosarito is the preferred route. Agua Higuera spring and cattle ranch are located about 1 kilometer up the wash to the right. The road then continues on to Misión San Borja (*See 4:52.5*).

45   This side road also leads to Misión San Borja. Both this road and the one at Kilometer 41.3 meet at Agua Higuera spring. The vegetation of this area is typical of that commonly found growing on similar flat areas along this section of the highway. Ocotillo, Cardon, Mesquite, Datilillo, two species of Agave, Jumping Cholla, Old Man Cactus, Cirio, and several species of annual composites are growing at the foot of the slopes. On the higher, drier parts of the slopes, Elephant Trees, Teddy Bear Cholla, Jumping Cholla, and Cirios grow in abundance. The vegetation growing along the pavement edge is Cheesebush, Brittlebush, Nightshade, and Desert Hollyhock.

48.5   The Elephant Tree on the left is covered by parasitic "Witches Hair."

54   Along with the first view of the Gulf, the 75 kilometer long Isla Angel de la Guardia is visible to the northeast. It forms the line of hills in the gap in the hills to the left. The highway descends along an alluviated stream course through pre-batholithic metasedimentary rocks.

56   The vegetation of this wash includes species characteristic of "Desert Wash Woodlands." It is predominated by Smoke Trees, Palo Verde, Ocotillo, Greasewood, Wax Plant, Ephedra, Cholla, a few large scattered Cardon and Cirio, and low annual scrubs.

58.7   The highway enters a canyon cut in the prebatholithic metasedimentary rocks. These metasedimentary rocks are intruded by numerous light colored dikes.

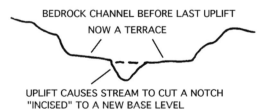

BEDROCK CHANNEL BEFORE LAST UPLIFT
NOW A TERRACE

UPLIFT CAUSES STREAM TO CUT A NOTCH
"INCISED" TO A NEW BASE LEVEL

58 To the right, below the highway, there is a gorge in the volcanics that was cut as an old stream meander was deepened by stream rejuvenation and uplift. Originally, the stream cut a surface on the rocks. As a result of uplift, the stream cut a gorge 10-15 feet deep leaving the former surface as a bedrock terrace on each side of the gorge.

61.5   The view is of the small, closer islands in the Bahía de Los Ángeles. Isla Angel de la Guardia dominates the center of the bay.

60.5   The highway drops onto an alluviated surface and turns to the south following the coastline to the main part of Bahía de Los Ángeles.

62.3   Bahía de Los Ángeles comes into view with the sandy point and lighthouse of Punta Arena visible to the left and the south point of Punta La Herradura (horseshoe) that forms the outer point of the bay.

The dry slopes around this horseshoe shaped bay are sparsely vegetated by Ocotillo, Palo Verde, Creosote Bush, and Elephant Trees.

66   **BAHÍA DE LOS ÁNGELES** is protected by the "midriff" islands producing a relatively calm bay loaded with corbina, yellowtail, sea bass, and other popular game fish.

Sunrise at Bahía de Los Ángeles

WHERE TO NOW? To leave Bahía de Los Ángeles, return the way you came or inquire locally about taking the 4-wheel dirt road leading southward through Las Flores, El Progresso, and El Arco to join the main highway between Guerrero Negro and San Ignacio.

REFERENCES

Allison, E.C., 1974, The type Alisitos Formation (Cretaceous, Aptian-Albian) of Baja California and its bivalve fauna *in* Gastil, G., and Lillegraven, J., eds, Guidebook; the geology of peninsular California: L.A, CA., Pac. Sec., Am. Assoc. of Petrol. Geol., p. 20-59.

Anderson, C.A., 1950, 1940 E.W. Scripps cruise to the Gulf of California, Part I: Geology of islands and neighboring land areas: Geological Society of America Memoir, 43, 53 p.

Ashby, J.R., Ku, T.L., & Minch, J.A., 1987, Uranium-series Ages of Corals from the Upper Pleistocene Mulegé Terrace, Baja California Sur, Mexico, Ciencias Marinas, v. 15, p. 139-141.

Ashby, J.R., & Minch, J.A., 1987, Late Tertiary Sedimentation and Molluscan Paleoecology of the Mulegé Embayment, Baja California Sur, Mexico, Geology, v. 15, p. 139-141.

Axelrod (1979),

Beal, C.H., 1948, Reconnaissance of the geology and oil possibilities of Baja California, Mexico: Geol. Soc. of Am., Mem., 31, p. 138.

Borcherdt, 1975 - fault diagram

Carreño, A.L., J. Ledesma-Vázquez y R. Guerrero-Arenas, 2000, Biostratigraphy And Depositional History Of The Tepetate Formation At Arroyo Colorado (Early-Middle Eocene), Baja California Sur, Mexico; Ciencias Marinas

Case, T.J., and Cody, M.L., 1983, Island Biogeography *in* the Sea of Cortez, Univ. of Calif. Press, Berkley, 508 p.

Duffield, W.A., 1968, Petrology and structure of the El Pinal Tonalite, Baja California, Mexico: Geological Society of America Bulletin, v. 79, p. 1351-1374.

Fife, D.L., Minch, J.A., and Crampton, P.L., 1967, A Late Jurassic Age for the Santiago Peak Volcanics, Geological Society of America Bull., v. 78, p. 299-304.

Fife, D.L., 1968, Geology of the Bahia Santa Rosalía Quadrangle, Baja California, Mexico [Master Thesis]: San Diego, CA, San Diego State College, 100 p.

Forman, J.A., Burke, W.H., Jr., Minch, J.A., and Yeats, R.S., 1971, Age of the Basement Rocks at Magdalena Bay, Baja California, Mexico, Sixty-seventh annual meeting, Cordilleran Section, Geol. Soc. of America, Riverside, CA.

Gastil, G., Minch, J., and Phillips, R.P., 1983, The geology and ages of the islands *in* Case, T.J., and Cody, M.L., editors, Island biogeography in the Sea of Cortez: Berkeley, CA, University of California Press, p. 13-25.

Gastil, R.G., Phillips, R.P., and Allison, E.C., 1975, Reconnaissance geology of the State of Baja California: Geol. Society of America Memoir, 140, 170 p.

Gastil, R.G., Krummenacher, D., and Minch, J., 1979, The record of Cenozoic volcanism around the Gulf of California: Geological Society of America Bulletin, v. 90, p. 839-857.

Harris, M.E., 1991, Geology of Baja California: A Bibliography, San Diego State Univ. Library

Hawkins, J.W., 1970, Petrology and possible tectonic significance of late Cenozoic volcanic rocks, southern California and Baja California: Geol. Soc. of Amer. Bull., v. 81, p. 3323-3338.

Heim, A., 1922, The Tertiary of southern Lower California: Geological Magazine, v. 59, p. 529-547.

Kilmer, F.H., 1963, Cretaceous and Cenozaic stratigraphy and paleontology, El Rosario area, Baja California, Mexico [Ph.D. dissertation]: Univ. of Calif., Berkeley, 216 p. 203.

Kilmer, F.H., 1965, A Miocene dugongid from Baja California, Mexico: Southern California Academy of Sciences Bulletin, v. 64, p. 57-74.

Kerstitch, Alex and Bertsch, Hans, Sea of Cortez Marine Invertebrates - 2nd Edition (Revised)

Knappe, R., Jr., 1974, The micropaleontology of a section of the Tepetate Formation, southern Baja California, and a paleobiogeographic comparison with equivalent foraminifera along the west coast of the United States [Master's thesis]: Athens, OH, Ohio University.

Lothringer, C.J., 1984, Geology of a Lower Ordovician Allochthon, Rancho San Marcos, Baja California, Mexico: *In* Frizzell, V.A., Jr., editor, Geology of the Baja California Peninsula: Pacific Section SEPM, p.17-22.

Luyendyk, B.P., Gans P.B., and Kamerling, M.J., 1988, $^{40}Ar/^{39}Ar$ Geochronology of Southern California Neogene Volcanism, *In* Weigand, P. W., ed., 1998, Contributions to the geology of the Northern Channel Islands, Southern California: American Assoc. of Petroleum Geologists, Pacific Section, MP 45.

Mallory, V.S., 1959, Lower Tertiary biostratigraphy of the California Coast Ranges, American Asso. Petrol. Geologists, Tulsa, 416p.

McCloy, C., 1984, Stratigraphy and depositional history of the San José del Cabo trough, Baja California Sur, Mexico. *In* Frizell, V.A., *ed.* Geology of the Baja California Peninsula, Pac. Sec. SEPM, pp. 267-273.

McLean, H., 1988, Reconnaissance geologic map of the Loreto and part of the San Javier quadrangles, Baja California Sur, Mexico: U.S. Geol. Survey Misc. Field Studies Map, Rpt # MF-2000, 10 p., geol. map 1:50,000.

McLean, H., 1989, Reconnaissance geology of a Pliocene marine embayment near Loreto, Baja California Sur, Mexico *in* Abbott, P.L., ed, Geologic studies in Baja California: Los Angeles, CA, Pac. Sec. SEPM, 63, p. 17-25.

Mina U.F., 1957, Bosquezo geologico del Territorio Sur de la Baja California: Asociacion Mexicana de Geologos Petroleros Boletin, v. 9(3-4), p. 141-269.

Minch, J.A., 1967, Stratigraphy and structure of the Tijuana-Rosarito Beach area, northwestern Baja California, Mexico: Geol. Society of America Bulletin, v. 78, p. 1155-1177.

Minch, J.A., 1971, Landsliding and the Effects on Resort Development Between Tijuana and Ensenada, Baja California, Mexico: *in* Coastal Studies in Baja California: Off. of Naval Res., Tech. Rpt. # 0-72-1, NR 387-045, 17 p.

Minch, J.A., 1972, The Late Mesozoic-Early Tertiary framework of continental sedimentation of the northern Peninsular Ranges, Baja California, Mexico: Doctoral, Univ. of Calif., Riverside, Riverside, California, p. 192.

Minch, J.A., Schulte, K.C., and Hofman, G., 1970, A Middle Miocene age for the Rosarito Beach Formation in northwestern Baja California, Mexico: Geological Society of America Bulletin, v. 81, p. 3149-3153.

Minch, J.A., Ashby, J.R., Demere, T.A., and Kuper, H.T., 1984, Correlational and depositional environments of the Middle Miocene Rosarito Beach Formation of northwestern Baja California, Mexico *in* Minch, J.A., and Ashby, J.R., editors, Miocene and Cretaceous depositional environments, northwestern Baja California, Mexico; annual meeting: Pac. Sec., Am. Assoc. of Petrol. Geol., 54, p. 33-46.

Morris, W.J., 1966, Fossil Mammals From Baja California, New Evidence on Early Tertiary Migrations: Science, v. 153, p. 1376-1378.

Morris, W.J., 1967, Baja California - Late Cretaceous dinosaurs: Science, v. 155, p. 1539-1541.

Morris, W.J., 1969, Late Cretaceous dinosaurs from Baja California: Geol. Soc. of Amer. Sp. Paper, 121, p. 209.

Morris, W.J., 1971, Mesozoic and Tertiary vertebrates in Baja California: National Geographic Society Research Reports, 1965, p. 195-198.

Normark, W.R., and Curray, J.R., 1968, Geology and structure of the tip of Baja California, Mexico: Geol. Soc. of America Bulletin, v. 79, p. 1589-1600.

Radamaker, K., 1995, ABA Field List of the Birds of Baja California.

Robbins, C.S., Brunn, B., and Zim, H.S., 1983, A Guide to Field Identification Birds of North America, Western Publishing Co., Inc., Racine.

Rowland, R.W., 1972, Paleontology and paleoecology of the San Diego Formation in northwestern Baja California: San Diego Soc. of Nat. History Trans., v.17, p.25-32.

Roberts, N.C., 1989, Baja California Plant Field Guide, Natural History Publishing Company, La Jolla.

Rebman, J., 2012, Baja California: Plant Guide, Sunbelt Publications, San Diego, 3rd edition, 480p.

Santillan, M., and Barrera, T., 1930, Las Posibilidades petroliferas en las Costa Occidental de Baja California, entre los paralelos 30 y 32 de latitud norte: Mexico (City) Universidad p. 1-37.

Scott, S.L., 1992, National Geographic Society Field Guide to the Birds of North America, National Geographic Society, Washington D.C.

Shreve, F., and Wiggins, I.L., 1964, Vegetation and Flora of the Sonoran Desert, Stanford University Press, Stanford.

Standley, P.C., 1920-1926, Contributions from the United States National Herbarium: Trees and Shrubs of Mexico, volume 23, GPO, Washington D.C.

White, C.A., 1885, On new Cretaceous fossils from California: U.S. Geol. Survey Bull. 22, p. 355-373.

Wiggins, I.L., 1980, Flora of Baja California, Stanford University Press, Los Angeles.

Wilson, I.F., 1948, Buried topography, initial structure, and sedimentation in Santa Rosalía area, Baja California, Mex.: Am. Assoc. of Petrol. Geol. Bull., v.32, p.1762-1807.

Wilson, I.F., and Rocha Moreno, V.S., 1957, Geology and mineral deposits of the Boleo copper district, Baja California, Mexico: USGS Prof. Paper 273, 134 p.

Woodford, A.O., 1928, The San Quintín volcanic field, Lower California: Am. Jour. of Sci., v. 215, 5th, v. 15, p. 337-345.

Yeats, R.S., Minch, J.A., and Forman, J.A., 1971, Paired basement terranes in Baja California Sur, Mexico: Geol. Soc. America Abstracts with Programs, v.3, p.760.

Other Popular Publications

Adams and Wyckoff: A Golden Guide to Landforms.

Zim and Shaffer: A Golden Guide to Rocks and Minerals.

Baja Explorer Topographic Atlas Directory

Crosby, H.W., 1997: Cave Paintings of Baja California, Discovering the Great Murals of an unknown People

Automobile Club Guide to Baja California.

Gohier, F.: A Pod of Grey Whales.

Gotshall, Daniel: Guide to Marine Invertebrates.

Peterson Field Guide, Pacific Coast Shells.

Peterson Field Guide, Western Birds.

Potter, Ginger: Baja Book IV.

GLOSSARY

AEOLIAN - Deposited by wind.

ALLUVIAL FANS - Alluvium piled in a fan, by streams, at the foot of a mountain in a desert region.

ALLUVIUM - Sand, gravel, silt currently being moved by and deposited in streams.

ANGULAR UNCONFORMITY - Unconformity in which the lower sedimentary beds are tilted and eroded before the upper sedimentary beds are deposited.

ANTICLINE - Fold where the beds are bowed up in the center.

AQUIFER - Water bearing layer, usually sandstone. 2:170.2, 8:212

ARKOSIC - Sand containing abundant feldspar. Usually formed in an arid climate.

BAJADA - Coalesced (joined together) alluvial fans often surrounding a playa.

BASE LEVEL - The lowest point to which a stream can erode.

BASE LEVELED - Eroded to nearly flat plain at or near base level.

BATHOLITH - Body of plutonic rock greater than 40 mi^2.

BIOLOGICAL BEACH - Beach formed from the calcareous remains of marine shells and other organisms. 6:115.8

BIOTURBATION - Mixing of sediments by burrowing animals.

CADODE - Stem or trunk which acts as a leaf to perform photosynthesis.

CALCAREOUS - Containing calcium carbonate.

CALICHE - Calcium Carbonate deposited just above the water table as groundwater, that is drawn upward by capillary action, evaporates.

CHEMICAL WEATHERING - Chemical break down of the minerals in a rock = decomposition; common in the tropics.

CONTACT - Boundary between rock units.

COQUINA - Conglomerates formed from shells.

DEFLATION - The removal of material from a surface by wind.

DETRITUS - Fragments of material which removed from their source by erosion.

DIKE - Igneous material filling a crack/joint in rocks. 2:90, 15:64

DIP - Angle of inclination of layers or faults. 3:113

ECOTONE - An ecological community of mixed vegetation formed by the overlapping of adjoining plant communities.

EDAPHIC - The way in which the soil affects living organisms.

EMBAYMENT - Indentation in a shoreline forming an open bay; area which contains rocks deposited in an embayment.

ENDEMIC - Native or confined to a certain region. 3:58

EPIPHYTIC - A plant that lives on another plant upon which it depends for mechanical support, but not for nutrients. 4:49

EROSION SURFACE - Relatively flat surface that has been base-leveled by streams.

ESCARPMENT - Steep cliff or line of cliffs typically along a fault.

ESTUARY - Where a river empties into the ocean through a tidal area. 1:69.2, 1:73

FACIES - Different rock units deposited at the same time in different parts of a basin.

FANGLOMERATE - Coarse sediments deposited in an alluvial fan.

FAULT - Fracture in rocks with movement.

FAULT SCARP - Uplifted cliff or bank along a fault line. 3:265,

FAULT ZONE - Most major faults are a series of smaller faults and related features. 2:24, 2:45.1, 12:147, 12:130

FLUVIAL - Sediments deposited by stream action.

FOLIATED - Alignment of mineral grains.

FORMATION - Mapable rock unit.

GEOSYNCLINE - Syncline on a continental scale. 5:162

GRABEN - Down dropped block bounded by faults. 9:89
HALOPHYTE - Salt tolerant plants.
HOGBACK - Narrow ridge with steeply dipping beds.
HORST - Uplifted block bounded by faults. 2:151.5
HOT SPRING - Spring in which the water temperature is $10°$ warmer than the mean air temperature. Usually found along faults. 2:18.5, 13:4.4
HYPABYSSAL - Near surface intrusive.
INDICATOR SPECIES - A plant restricted to, and indicates a specific environment.
IGNEOUS - Formed by cooling of molten material. *See* Rock Chart.
INTRUSION - The injection of molten material into an area -below the surface.
ISOTOPIC AGE - Probable age determined by radiometric age dating.
JOINT - Fracture without movement.
K/AR - Potassium 40 decays to Argon 40 at a predictable rate. Heat and weathering can affect the dates obtained. (See Radiometric Age Dating)
LACUSTRINE - Sediments deposited in a lake. 3:125
LAHAR - Andesite flow that broke up as it moved downslope.
LAPILLI - Coarse grained volcanic ejecta.
LATERAL FAULT (STRIKE SLIP FAULT) - Fault with largely horizontal motion with one side slipping past the other.
LICHENS - An algae and a fungus got together and took a lichen to each other. The algae provides the food and the fungus the shelter.
LITHIFICATION - The process by which sediment is converted to sedimentary rock (cementation, compaction, desiccation, crystallization).
LITTORAL - Shoreline between tides.
MAGMA - Molten rock formed by release of pressure on ocean ridges/rifts or by friction in subduction zones. *See* Rock Chart.
MATRIX - Material around the grains in a rock.
MECHANICAL WEATHERING - Disintegration; mechanical break down - more common in deserts.
MEGA-BRECCIA - Large clasts in breccia.
MESIC - Moist conditions.
METAMORPHIC - Rocks which have undergone change by being subjected to heat (not melting) and pressure under the surface of the earth.
METAMORPHISM - Application of heat and pressure to change the form of minerals in a rock.
METASEDIMENTARY - Sedimentary rocks which have undergone low grade metamorphism.
METAVOLCANIC ROCKS - Volcanic rocks which have undergone low grade metamorphism. 3:113
MONOCLINE - Local steepening of beds, usually over a fault. 1:14.8
NONCONFORMITY - Sedimentary beds deposited on the eroded surface of igneous or metamorphic rocks.
NORMAL FAULT - Caused by tension where the overhanging block slides downward. Main fault of horst and graben areas.
ONYX (TRAVERTINE) - Banded calcite formed in hot spring deposits.
OPPORTUNISTIC - An organism taking advantage of opportunities for self advancement.
PARASITE - An organism that feeds and grows on a host. 4:49
PEDIMENT - Gently sloping surface cut by streams, in an arid climate, at the foot of a mountain range. Grades into an alluvial fan.
PEGMATITE - Coarse grained dike, often with rare minerals such as beryl, tourmaline, or topaz.

PENEPLAIN - Low relief erosion surface cut by streams in a humid climate.

PHYTOGEOGRAPHIC - The grouping of plants by geographic occurrence.

PLATEAU - Extensive area of volcanic flows, usually basalt.

PLAYA /PLAYA LAKE - Desert area of interior drainage where water and fine sediments accumulate. 16:33

PLUGS - Eroded necks of volcanoes. 1:39

PLUTON - Individual body of plutonic rock.

PORPHYRY - Igneous rock with larger crystals in a finer grained matrix. *See* Rock Chart.

PREBATHOLITHIC - Rocks formed prior to the intrusion of the Batholith.

PYROCLASTIC - Fragmented volcanic material ejected into the air from a volcano.

RADIOMETRIC AGE DATING - Certain elements change from a parent element to a daughter element. This change is random. However, the rate is predictable with large numbers. This rate is called a half-life which is the time for 1/2 of the parent to decay to the daughter; 1, 1/2, 1/4, 1/8, 1/16. The ratio determines the age.

REVERSE FAULT / THRUST FAULT - Hanging wall block is pushed up and over the footwall block. Formed by compression.

SCARIFICATION - Abrasion of the coating on a seed by running water.

SEDIMENTS - Fragments of preexisting rocks worn from the land and deposited by wind, water, ice, or gravity.

SEDIMENTARY - Rocks made up of fragments of rocks- deposited by running water, wind, ice, and gravity, or by accumulation of various organic materials formed by chemical action.

SHUTTER RIDGES - Formed when a fault moves a ridge to block the drainage.

SLUMP - Rotational block slide.1:64

SPHEROIDAL WEATHERING - Igneous rocks have a tendency to weather into spheres. 3:162

SPIT - Strip of sand or mud, projecting from the shoreline. 6:182

STACK - Small isolated, near shore rock mass. 9:6

STRATA - Layering in sedimentary rocks due to changes in regimen.

STRATIGRAPHIC - Pertaining to the description of the strata.

STRIKE - Horizontal bearing of tilted layers or fault line.

STRIKE SLIP FAULT - A fault that has lateral movement.

SUBDUCTION ZONE - Where one plate dives beneath another along a convergent plate boundary.

SUBMARINE FAN - Sediment moving down a submarine canyon forms a fan shaped mass at the end of the canyon.

SUBSTRATE - Rocks or sediments at the land/water interface.

SYNCLINE - Downfold or basin where beds drop towards each other.

TALUS - Landslide where the movement is a few rocks at a time.

TÓMBOLO - Strip of sand or mud, deposited by waves or currents, connecting an island with the mainland. 6:92

TUFF - Volcanic fragments ejected from volcano. TUFFACEOUS - Containing tuff

TURBIDITE - A turbid mass of muddy water moving down slope.

VOLCANIC - Fine grained igneous rock which cooled rapidly at or near the surface.

VOLCANICLASTIC - Fragments of volcanic rock.

XERIC - Dry conditions.

XEROPHYTE - Plants adapted to dry conditions. 3:86

INDEX

This INDEX lists topics elaborated on in the text including Phytogeographic Areas, Avifauna, and photos of plants.
2-34 refers to Kilometer 34 of Chapter 2. Bio denotes Biology section. This is not a comprehensive index.